PARTICLES AND SOURCES

DOCUMENTS ON MODERN PHYSICS

Edited by

Elliott W. Montroll, University of Rochester
George H. Vineyard, Brookhaven National Laboratory
Maurice Lévy, Université de Paris

Additional Volumes in preparation

Particles and Sources

JULIAN SCHWINGER

Harvard University

Notes by Tung-mow Yan

GORDON AND BREACH, SCIENCE PUBLISHERS

New York   *London*   *Paris*

Library of Congress Catalog Card Number: 69–19390

Editorial office for Great Britain:

Gordon and Breach, Science Publishers Ltd.
8 Bloomsbury Way
London W.C. 1

Editorial office for France:

Gordon & Breach
7–9, rue Emile Dubois
Paris 14e

Distributed in Canada by:

The Ryerson Press
299 Queen Street West
Toronto 2B, Ontario

(0206)

Printed in Switzerland by City-Druck AG, Zurich

Editors' Preface

Seventy years ago when the fraternity of physicists was smaller than the audience at a weekly physics colloquium in a major university, a J. Willard Gibbs could, after ten years of thought, summarize his ideas on a subject in a few monumental papers or in a classic treatise. His competition did not intimidate him into a muddled correspondence with his favorite editor nor did it occur to his colleagues that their own progress was retarded by his leisurely publication schedule.

Today the dramatic phase of a new branch of physics spans less than a decade and subsides before the definitive treatise is published. Moreover, modern physics is an extremely interconnected discipline and the busy practitioner of one of its branches must be kept aware of breakthroughs in other areas. An expository literature which is clear and timely is needed to relieve him of the burden of wading through tentative and hastily written papers scattered in many journals.

To this end we have undertaken the editing of a new series, entitled *Documents on Modern Physics*, which will make available selected reviews, lecture notes, conference proceedings, and important collections of papers in branches of physics of special current interest. Complete coverage of a field will not be a primary aim. Rather, we will emphasize readability, speed of publication, and importance to students and research workers. The books will appear in low-cost paper-covered editions, as well as in cloth covers. The scope will be broad, the style informal.

From time to time, older branches of physics come alive again, and forgotten writings acquire relevance to recent developments. We expect to make a number of such works available by including them in this series along with new works.

ELLIOTT W. MONTROLL
GEORGE H. VINEYARD
MAURICE LÉVY

Contents

PARTICLES AND SOURCES

1. NONINTERACTING PARTICLES

1.1 Introduction

Research in particle physics presently falls in two main classifications: Quantum Field Theory and S-Matrix Theory. While quantum field theory, like field theorists, comes in all sizes and shapes we regard its basic characteristics to be as follows. It is a space-time formulated operator theory. The fundamental dynamical variables, the fields, describe certain localized excitations, which in particle language, correspond to all possible combinations of particles with the prescribed quantum numbers. When the physical couplings are weak, as in electrodynamics, the relation between field and particle may appear to be close. For strongly interacting systems it is certainly very remote. Current algebra was originally motivated by a reaction against field theory, in which currents, supposedly more physical, were introduced as basic operators. But the latest development in this area has been a return to the phenomenological field as primary. The practical difficulty in connecting field-theory hypotheses with raw experimental data led to the phenomenological emphasis of S-matrix theory. Here the particle is primary. Attention is concentrated on the momentum space description of various collisions. Dynamical principles are sought in the analytic extension of the momenta to complex values.

In these lectures we shall describe a totally new approach to particle physics. It is intermediate in concept between the two previous formulations. It

1

shares with field theory the physical emphasis upon space and time, but it is not an operator theory. Like S-matrix theory, it is phenomenological in its emphasis upon the actual physical system, but there is no reference to analyticity in momentum space. The results of quantum electrodynamics are reproduced without the irrelevance of divergences, or renormalizations. This means that we remove the unnecessary field-theory hypothesis that detailed space-time description is possible, down to distances smaller than those involved in presently accessible phenomena. Characteristic S-matrix features such as dispersion relations are deduced, not from abstract mathematical requirements, but through intuitive space-time arguments. Current algebra results, which have lent credence to detailed speculations concerning the inner structures of particles, are recovered by purely phenomenological procedures. Thus, a unified method is now available, one that is economical in hypothesis and effective in application.

1.2   Sources

The concept of particle has been steadily extended in the last few decades. From the stable electron and proton, to the very long-lived neutron, to the short-lived π and Λ, to the highly unstable p and N^* it has progressed to more and more short-lived excitations. Thus, in general, particles must be created in order to study them, since most of them are unstable. In a general sense this is also true of high-energy stable particles, which must be created in that situation by some device, i.e. an accelerator. One can regard all such creation acts as collisions, in which the necessary properties are transferred from other particles to the one of interest. There will be a variety of collisions that can serve to create a particle with specified properties. The other particles in the collision appear only to supply these attributes. They are, in an abstract sense, the source of the particle in question. In other words, the source concept is the abstraction of all possible dynamical mechanisms whereby the particular particle can be produced. We try to represent this abstraction of realistic processes numerically. The fact that the collision will have some degree of space-time localizability indicates the utility of a numerical function $S(x)$ to measure this aspect. The effectiveness of the collision in liberating various momenta can be measured by another function $S(p)$. The quantum-mechanical requirement of complementarity can then be imposed, as in

$$S(p) = \int (dx)e^{-ipx}S(x), \qquad\qquad (1.1)$$

where

$$(dx) = dx^0 dx^1 dx^2 dx^3,$$

$$px = p_\mu x^\mu = \mathbf{p} \cdot \mathbf{x} - p^0 x^0,$$

and units $\hbar = c = 1$ are used.

Unstable particles eventually decay and the decay process is a detection device. More generally, any detection device can be regarded as removing or annihilating the particle. Thus the source concept can again be used as an abstraction of annihilation collisions, with the source acting negatively, as a sink. We now have a total picture of any physical situation in which sources are used to create the initial particles from the vacuum state, and sources are used to detect the final particles resulting from some interaction, and then we return to the vacuum state.

1.3   Spinless Particles

To make the concept the basis of a quantitative theory, we consider first a stable spinless neutral particle of mass m. To begin with, we use a weak source, i.e. one for which multiparticle production is negligible. To specify a weak source, we consider its effectiveness in creating a particle with momentum $\mathbf{p}$, in the small range $(d\mathbf{p})$. An invariant measure of momentum space is

$$d\omega_p = \frac{(d\mathbf{p})}{(2\pi)^3} \frac{1}{2p^0}, \qquad p^0 = +\sqrt{\mathbf{p}^2+m^2}. \tag{1.2}$$

We now define the source K in terms of the creation and annihilation probability amplitudes

$$\langle 1_p|0_-\rangle^K = \sqrt{d\omega_p}\, iK(p), \tag{1.3}$$

$$\langle 0_+|1_p\rangle^K = \sqrt{d\omega_p}\, iK(-p), \tag{1.4}$$

which convey the idea that the source liberates or absorbs momentum p in the respective processes. The square-root signs appear because the probabilities are proportional to the momentum range $(d\mathbf{p})$. The subscript on the vacuum state indicates the time sense, $|0_-\rangle$ is the vacuum state before the source has operated. The factors of i are included here for later convenience. There are probability requirements that express the weak source restriction:

$$1 \cong |\langle 0_+|0_-\rangle^K|^2 + \sum_p |\langle 1_p|0_-\rangle^K|^2, \tag{1.5}$$

$$1 \cong |\langle 0_+|0_-\rangle^K|^2 + \sum_p |\langle 0_+|1_p\rangle^K|^2. \tag{1.6}$$

Since

$$|K(p)|^2 = |\int (dx) e^{-ipx} K(x)|^2$$

$$= |\int (dx) e^{ipx} K^*(x)|^2$$

$$= |K^*(-p)|^2,$$

these two conditions, (1.5) and (1.6), will be equivalent if $K(x)$ is real. We shall take this to be a general property of any source; a complex source is regarded as a combination of two real sources which describes some physical property of that multiplicity. We also notice, from the restriction (1.5) or (1.6), that $|\langle 0_+|0_-\rangle^K|$ must differ from unity by terms of order K^2. It is consistent to assume that no terms of order K appear in $\langle 0_+|0_-\rangle^K$, as we shall verify later.

Now consider a complete situation in which particles are created by K_2, propagate in space and time, and then are detected by K_1 (Fig. 1.1). In this

FIGURE 1.1

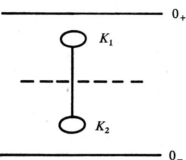

case the causal situation is well-defined, the detection source $K_1(x)$ is certainly localized in time later than the emission source $K_2(x)$. The overall description is then expressed by the vacuum probability amplitude

$$\langle 0_+|0_-\rangle^K \cong \langle 0_+|0_-\rangle^{K_1}\langle 0_+|0_-\rangle^{K_2} + \sum_p \langle 0_+|1_p\rangle^{K_1}\langle 1_p|0_-\rangle^{K_2}$$

$$\cong 1 + O(K_1{}^2) + O(K_2{}^2)$$

$$+ \int d\omega_p \int (dx)(dx') iK_1(x)e^{ip(x-x')} iK_2(x'). \tag{1.7}$$

We regard $K_1(x)$ and $K_2(x)$ as two different manifestations of the same physical mechanism, that is, they are the values of one general source in different space-time regions. Therefore the only possible combination that can occur is the total source

$$K = K_1 + K_2.$$

This is a fundamental postulate, the principle of the unity of the source, which embodies the idea of the uniformity of nature. Then the vacuum amplitude must have the general quadratic form

$$\langle 0_+|0_-\rangle^K = 1 + (i/2)\int (dx)(dx')K(x)\Delta_+(x-x')K(x'), \tag{1.8}$$

where it is known that

$$x^0 > x^{0'} : (1/2)[\Delta_+(x-x') + \Delta_+(x'-x)] = i\int d\omega_p e^{ip(x-x')}. \tag{1.9}$$

Since only the part of $\Delta_+(x-x')$ that is symmetrical in x and x' can contribute to the integral, we are permitted to define

$$\Delta_+(x-x') = \Delta_+(x'-x), \tag{1.10}$$

and

$$x^0 > x^{0'} : \Delta_+(x-x') = i\int d\omega_p e^{ip(x-x')}. \tag{1.11}$$

This function is familiar. It is the solution of the differential equation

$$(m^2 - \partial^2)\varDelta_+(x-x') = \delta(x-x') \tag{1.12}$$

that obeys the boundary condition of outgoing waves in time, that is, positive (negative) frequencies for positive (negative) time differences. It should be remarked that we did not begin with these requirements, they appear as a consequence of our theory. We record here the alternative representation for $\varDelta_+(x-x')$,

$$\varDelta_+(x-x') = \int \frac{(dp)}{(2\pi)^4} \frac{e^{ip(x-x')}}{p^2 + m^2 - i\varepsilon}\bigg|_{\varepsilon \to +0}. \tag{1.13}$$

The quadratic terms in K_1 and K_2 that appear in $\langle 0_+|0_-\rangle^K$ now reproduce the structure of the product of $\langle 0_+|0_-\rangle^{K_1}$ and $\langle 0_+|0_-\rangle^{K_2}$. We can now proceed to check the probability requirement, (1.5) or (1.6). From (1.8), we get

$$|\langle 0_+|0_-\rangle^K|^2 = 1 - \int (dx)(dx')K(x) \, \mathrm{Re} \, (1/i)\varDelta_+(x-x')K(x') \tag{1.14}$$

for a weak source. The notation "Re" means "the real part of". Now

$$\mathrm{Re}(1/i)\varDelta_+(x-x') = \mathrm{Re} \int d\omega_p e^{ip(x-x')}, \tag{1.15}$$

which holds for all $x-x'$, since the right-hand side is symmetrical in x and x'. Thus

$$\begin{aligned}
\int (dx)(dx')K(x) \, \mathrm{Re} \, (1/i)\varDelta_+(x-x')K(x') \\
= \mathrm{Re} \int d\omega_p K(-p)K(p) \\
= \int d\omega_p |K(p)|^2,
\end{aligned} \tag{1.16}$$

and (1.14) becomes

$$|\langle 0_+|0_-\rangle^K|^2 = 1 - \sum_p |\langle 1_p|0_-\rangle^K|^2, \tag{1.17}$$

which is equivalent to (1.5) or (1.6). It can now be recognized that the physically necessary minus sign in (1.17) is the consequence of the particular way in which the factors of i in $\langle 1_p|0_-\rangle^K$ and $\langle 0_+|1_p\rangle^K$ have been introduced. These discussions show that the formalism is internally consistent.

To remove the restriction to weak sources, we make use of the possibility of preparing directional sources to arrange an arbitrary number of pairs of weak emission and weak absorption sources in such a way that a particle emitted by the emission source of one pair will not be detected by the detection source of another. This situation is illustrated in Fig. 1.2. Since these processes are physically independent, we have simply

$$\langle 0_+|0_-\rangle^K = \prod_\alpha \left[1 + (i/2)\int (dx)(dx')K_\alpha(x)\varDelta_+(x-x')K_\alpha(x') \right], \tag{1.18}$$

FIGURE 1.2

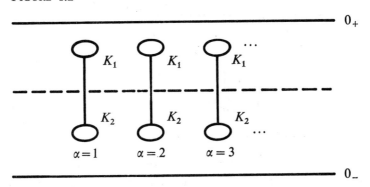

where

$$K_\alpha = (K_1 + K_2)_\alpha ,$$

refers to a particular source pair. Insisting again upon the unity of sources, we require that only the combination

$$K = \sum_\alpha K_\alpha \tag{1.19}$$

occurs. Since

$$\int (dx)(dx')K_\alpha(x)\varDelta_+(x-x')K_\beta(x') \cong 0 \text{ for } \alpha \neq \beta, \tag{1.20}$$

the appropriate expression is

$$\langle 0_+ | 0_- \rangle^K = \exp\left[(i/2)\int (dx)(dx')K(x)\varDelta_+(x-x')K(x') \right]. \tag{1.21}$$

Conversely, under the condition (1.20) and that the sources are weak, (1.21) reproduces (1.18), because

$$\exp\left[(i/2)\int (dx)(dx')K(x)\varDelta_+(x-x')K(x') \right]$$
$$\cong \exp\left[(i/2)\sum_\alpha \int (dx)(dx')K_\alpha(x)\varDelta_+(x-x')K_\alpha(x') \right]$$
$$= \prod_\alpha \exp\left[(i/2)\int (dx)(dx')K_\alpha(x)\varDelta_+(x-x')K_\alpha(x') \right]$$
$$\cong \prod_\alpha \left[1 + (i/2)\int (dx)(dx')K_\alpha(x)\varDelta_+(x-x')K_\alpha(x') \right].$$

We have produced, under special circumstances, a situation in which at a given time an arbitrary number of noninteracting particles can be present. Let's now extend this result to all analogous situations in which the physical interactions among the particles are not significant but the particles need not be macroscopically isolated, so that microscopic quantum

FIGURE 1.3

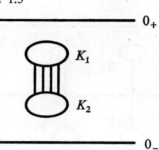

interference effects come into play. Thus consider more generally multiparticle exchanges between sources under the noninteracting causal condition (Fig.1.3)

$$K = K_1 + K_2.$$

Then

$$\langle 0_+ | 0_- \rangle^K = \langle 0_+ | 0_- \rangle^{K_1} \exp\left[i \int (dx)(dx') K_1(x) \Delta_+(x-x') K_2(x') \right]$$
$$\times \langle 0_+ | 0_- \rangle^{K_2}, \tag{1.22}$$

where, as before, keep in mind that $x^0 > x^{0'}$, and therefore

$$i \int (dx)(dx') K_1(x) \Delta_+(x-x') K_2(x')$$
$$= \int d\omega_p i K_1(-p) i K_2(p)$$
$$= \sum_p i K_{1p}{}^* i K_{2p} \tag{1.23}$$

in which we have defined

$$K_p = \sqrt{d\omega_p}\, K(p). \tag{1.24}$$

We now analyze the vacuum amplitude in terms of multiparticle states

$$\langle 0_+ | 0_- \rangle^K = \sum_{\{n\}} \langle 0_+ | \{n\} \rangle^{K_1} \langle \{n\} | 0_- \rangle^{K_2}. \tag{1.25}$$

This is accomplished by writing

$$\exp\left(\sum_p i K_{1p}{}^* i K_{2p}\right) = \prod_p \exp\left(i K_{1p}{}^* i K_{2p}\right)$$
$$= \prod_p \sum_{n_p=0}^{\infty} \frac{(i K_{1p}{}^*)^{n_p}}{\sqrt{n_p!}} \cdot \frac{(i K_{2p})^{n_p}}{\sqrt{n_p!}}. \tag{1.26}$$

Thus, we identify

$$\langle \{n\} | 0_- \rangle^K = \prod_p \frac{(i K_p)^{n_p}}{\sqrt{n_p!}} \langle 0_+ | 0_- \rangle^K, \{n\} = \{n_p\}, \tag{1.27}$$

and

$$\langle 0_+ | \{n\} \rangle^K = \langle 0_+ | 0_- \rangle^K \prod_p \frac{(iK_p^*)^{n_p}}{\sqrt{n_p!}}, \tag{1.28}$$

which clearly describe multiparticle states with $n_p = 0, 1, 2, \cdots$ particles in each momentum space cell. Since the particles are not individually distinguishable, and there is no limit to the number of particles with a specified property, we are describing identical particles obeying Bose–Einstein (BE) statistics.

As a check of consistency, let's test the probability normalization condition, or equivalently, the completeness of the multiparticle states,

$$\sum_{\{n\}} \langle 0_- | \{n\} \rangle^K \langle \{n\} | 0_- \rangle^K = 1, \tag{1.29}$$

$$\langle 0_- | \{n\} \rangle^K = [\langle \{n\} | 0_- \rangle^K]^*.$$

Inserting (1.27) and its complex conjugate, we get

$$\sum_{\{n\}} \langle 0_- | \{n\} \rangle^K \langle \{n\} | 0_- \rangle^K = |\langle 0_+ | 0_- \rangle^K|^2 \prod_p \exp(|K_p|^2)$$

$$= |\langle 0_+ | 0_- \rangle^K|^2 \exp(\sum_p |K_p|^2). \tag{1.30}$$

But direct evaluation gives

$$|\langle 0_+ | 0_- \rangle^K|^2 = \exp\left[-\int (dx)(dx') K(x) \,\text{Re}\,(1/i)\Delta_+(x-x') K(x')\right]$$

$$= \exp\left[-\int d\omega_p K(-p)K(p)\right],$$

i.e.

$$|\langle 0_+ | 0_- \rangle^K|^2 = \exp(-\sum_p |K_p|^2), \tag{1.31}$$

and the completeness is verified. Notice how the structure of $\langle 0_+ | 0_- \rangle^K$ enters in two different ways. On the one hand it gives $|\langle 0_+ | 0_- \rangle^K|^2$ directly, and on the other it is used to generate $\langle \{n\} | 0_- \rangle^K / \langle 0_+ | 0_- \rangle^K$. The consistency of these two procedures serves as a severe test of the formulation.

1.4   An Application. Stimulated Emission

The original definition of a creation source refers to the creation of particles from the vacuum state. In connection with the identification of BE identical particles, it is of some interest to generalize the meaning of a creation source to a situation in which particles may be present initially, as illustrated in Fig. 1.4. In fact, the general formula (1.21) will supply the answer to this problem. Let the initial particles be created by a creation

FIGURE 1.4

FIGURE 1.5

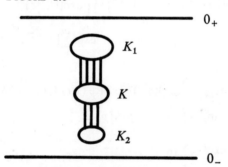

source K_2, and the final particles be detected by a detection source K_1. The well-defined causal arrangement is indicated in Fig. 1.5. The total source, now denoted by (K), is the sum of its constituents

$$(K) = K_1 + K_2 + K.$$

Now (1.21) gives

$$\langle 0_+ | 0_- \rangle^{(K)} = \langle 0_+ | 0_- \rangle^{K_1 + K_2} \langle 0_+ | 0_- \rangle^K$$

$$\times \exp\{i\int (dx)(dx')[K_1(x)\Delta_+(x-x')K(x')$$

$$+ K(x)\Delta_+(x-x')K_2(x')]\}$$

$$= \sum_{\{n\}} \langle 0_+ | \{n\} \rangle^{K_1} \langle \{n\} | 0_- \rangle^{K_2} \langle 0_+ | 0_- \rangle^K$$

$$\times \exp\Big[\sum_p (iK_{1p}{}^* iK_p + iK_p{}^* iK_{2p}) \Big]. \tag{1.32}$$

If K is a weak source, only terms linear in K are significant, and

$$\langle 0_+|0_-\rangle^{(K)} = \sum_{\{n\}}\langle 0_+|\{n\}\rangle^{K_1}\langle\{n\}|0_-\rangle^{K_2}$$

$$\times\left[1+\sum_p(iK_{1p}{}^*iK_p+iK_p{}^*iK_{2p})\right]. \tag{1.33}$$

But

$$iK_p\langle\{n\}|0_-\rangle^K = \sqrt{n_p+1}\,\langle\{n\}+1_p|0_-\rangle^K, \tag{1.34}$$

$$\langle 0_+|\{n\}\rangle^K iK_p{}^* = \langle 0_+|\{n\}+1_p\rangle^K\sqrt{n_p+1}, \tag{1.35}$$

which follow from (1.27) and (1.28), so that the terms linear in K are

$$\sum_{\{n\}}\sum_p\Big[\langle 0_+|\{n\}+1_p\rangle^{K_1}\sqrt{n_p+1}\,iK_p\langle\{n\}|0_-\rangle^{K_2}$$

$$+\langle 0_+|\{n\}\rangle^{K_1}\sqrt{n_p+1}\,iK_p{}^*\langle\{n\}+1_p|0_-\rangle^{K_2}\Big].$$

On the other hand, we have the general relation

$$\langle 0_+|0_-\rangle^{(K)} = \sum_{\{n\},\{n'\}}\langle 0_+|\{n\}\rangle^{K_1}\langle\{n\}|\{n'\}\rangle^K\langle\{n'\}|0_-\rangle^{K_2}.$$

On comparison, we infer the matrix elements

$$\langle\{n\}+1_p|\{n\}\rangle^K = \sqrt{n_p+1}\,iK_p, \tag{1.36}$$

and

$$\langle\{n\}|\{n\}+1_p\rangle^K = \sqrt{n_p+1}\,iK_p{}^*. \tag{1.37}$$

(1.36) can be rewritten as

$$\langle\{n\}+1_p|\{n\}\rangle^K = \sqrt{n_p+1}\langle 1_p|0_-\rangle^K \tag{1.38}$$

which gives the characteristic BE connection of spontaneous and stimulated emission, the total emission probability being proportional to n_p+1, with n_p the number of particles of momentum p present initially.

1.5   Spin-1 Particles. The Photon

The function $\Delta_+(x-x')$ is an invariant one which depends only on $(x-x')^2$. Thus the fundamental structure for noninteracting spin-0 particles

$$\langle 0_+|0_-\rangle^K = \exp\left[(i/2)\int (dx)(dx')K(x)\Delta_+(x-x')K(x')\right]$$

has an invariant significance, if $K(x)$ is transformed as a scalar. This is characteristic of spin-0 particles. The exponential quadratic form is, in fact, entirely general, although it is derived within the context of spin-0 particles. It simply describes the creation of particles, propagation of particles in space and time, and the detection of particles. The extra multiplicities that the particle possesses will superimpose on this basic structure. As a first step to remove the spin restriction, suppose the scalar is replaced by a vector $J_\mu(x)$, and consider the provisional structure

$$\langle 0_+|0_-\rangle^J = \exp\left[(i/2)\int (dx)(dx')J^\mu(x)\Delta_+(x-x')J_\mu(x')\right]. \tag{1.39}$$

However, this by itself will violate physical requirements, for the vacuum persistence probability implied by (1.39) is

$$\begin{aligned}|\langle 0_+|0_-\rangle^J|^2 &= \exp\left[-\int (dx)(dx')J^\mu(x)\,\text{Re}\,(1/i)\Delta_+(x-x')J_\mu(x')\right]\\ &= \exp\left[-\int d\omega_p J^\mu(-p)J_\mu(p)\right]\\ &= \exp\left[-\int d\omega_p\{|\mathbf{J}(p)|^2-|J^0(p)|^2\}\right],\end{aligned} \tag{1.40}$$

which is not necessarily less than or equal to unity, as expressed by the positiveness requirement

$$\int d\omega_p\left[|\mathbf{J}(p)|^2-|J^0(p)|^2\right]\geq 0.$$

Therefore, the time component of J_μ, which appears with the wrong sign, must be suppressed, in an invariant manner. This can be accomplished by the observation that the time-like momentum p^μ $(-p^2=m^2>0)$ supplies a natural time-like direction. On writing

$$J^\mu(p)^*J_\mu(p) = J^\mu(p)^*g_{\mu\nu}J^\nu(p)$$

we now replace the metric tensor $g_{\mu\nu}$ by the combination

$$g_{\mu\nu}+\frac{p_\mu p_\nu}{m^2}$$

which, in the rest frame of $p(\mathbf{p}=0, p^0=m)$, takes the following values

$$g_{\mu\nu}+\frac{p_\mu p_\nu}{m^2} = \begin{cases} \mu,\nu=k,l & :\delta_{kl} \\ \mu,\nu=0,k \text{ or } k,0: 0 \\ \mu,\nu=0,0 & :0 \end{cases} \tag{1.41}$$

and

$$J^\mu(p)^* \left(g_{\mu\nu} + \frac{p_\mu p_\nu}{m^2} \right) J^\nu(p) = |\mathbf{J}(p)|^2 \geqslant 0. \tag{1.42}$$

Thus, the correct structure, in coordinate space, is

$$\langle 0_+ | 0_- \rangle^J = \exp\left[\frac{i}{2} \int (dx)(dx') J^\mu(x) \left(g_{\mu\nu} \right.\right.$$
$$\left.\left. - \frac{1}{m^2} \partial_\mu \partial_\nu \right) \Delta_+(x-x') J^\nu(x') \right]. \tag{1.43}$$

Insistence upon the physical positiveness property has left us, for each momentum in its rest frame, with three independent sources, which are transformed into each other by spatial rotations. We are clearly describing a spin-1 particle. In an arbitrary coordinate frame we have

$$g_{\mu\nu} + \frac{1}{m^2} p_\mu p_\nu = \sum_{\lambda=1}^{3} e_{\mu p\lambda} e_{\nu p\lambda}^*. \tag{1.44}$$

Here

$$p^\mu e_{\mu p\lambda} = 0, \ \lambda = 1, 2, 3. \tag{1.45}$$

$$e^\mu_{p\lambda}{}^* e_{\mu p\lambda'} = \delta_{\lambda\lambda'}, \ \lambda, \lambda' = 1, 2, 3. \tag{1.46}$$

where $e_{p\lambda}$ are the polarization vectors associated with momentum p^μ, which may be taken to be real. As a check of dimensionality, set $\mu = \nu$ in (1.44) and sum over μ. We get

$$4 - 1 = 3,$$

which gives the right account of multiplicity. With the definition

$$J_{p\lambda} = \sqrt{d\omega_p} \, e_{\mu p\lambda}{}^* J^\mu(p), \tag{1.47}$$

we have

$$|\langle 0_+ | 0_- \rangle^J|^2 = \exp\left(-\sum_{p\lambda} |J_{p\lambda}|^2 \right). \tag{1.48}$$

As this expression suggests, the whole story of spin-0, such as identification of multiparticle states, check of completeness, etc., can be repeated here with λ added to p as quantum numbers. For instance

$$\langle \{n\} | 0_- \rangle^J = \prod_{p\lambda} \frac{(iJ_{p\lambda})^{n_{p\lambda}}}{\sqrt{n_{p\lambda}!}} \langle 0_+ | 0_- \rangle^J, \tag{1.49}$$

and

$$\langle 0_+ | \{n\} \rangle^J = \langle 0_+ | 0_- \rangle^J \prod_{p\lambda} \frac{(iJ_{p\lambda}{}^*)^{n_{p\lambda}}}{\sqrt{n_{p\lambda}!}}. \tag{1.50}$$

Since this discussion depends heavily on the fact that $m \neq 0$, it naturally raises a question concerning $m = 0$, where $g_{\mu\nu} - \dfrac{1}{m^2} \partial_\mu \partial_\nu$ is no longer meaningful. Let's return to the original expression (1.40). The only alternative now is to restrict the source in a covariant manner so that $|\mathbf{J}(p)|^2 - |J^0(p)|^2 \geqslant 0$. In particular, $\mathbf{J}(p) = 0$ must imply $J^0(p) = 0$. Since we are looking for a scalar restriction among the components of J^μ, and there is only one vector p^μ available, the only possibility is the scalar equation

$$p^\mu J_\mu(p) = \mathbf{p} \cdot \mathbf{J}(p) - p^0 J^0(p) = 0, \tag{1.51}$$

or, in coordinate space,

$$\partial_\mu J^\mu(x) = 0. \tag{1.52}$$

We are obviously describing the photon. Its masslessness demands that its source be a divergenceless or conserved vector. Incidentally, if the derivatives in (1.43) are transferred to the sources, the mass-dependent part has the structure $\dfrac{1}{m^2} \partial_\mu J^\mu(x) \varDelta_+(x - x') \partial_\nu J^\nu(x')$, which disappears if the source is restricted to be a conserved vector. And then one has a smooth transition from $m \neq 0$ to $m = 0$.

To see that the condition (1.52) is sufficient as well as necessary for the correct physical interpretation of

$$\langle 0_+ | 0_- \rangle^J = \exp\left[(i/2) \int (dx)(dx') J^\mu(x) D_+(x - x') J_\mu(x') \right], \tag{1.53}$$

where

$$D_+(x) = \varDelta_+(x, m^2 = 0), \tag{1.54}$$

we calculate the vacuum persistence probability

$$|\langle 0_+ | 0_- \rangle^J|^2 = \exp\left[- \int d\omega_p J^\mu(-p) g_{\mu\nu} J^\nu(p) \right], \tag{1.55}$$

by constructing a dyadic representation for $g_{\mu\nu}$. For $m \neq 0$, p^μ provides a time-like vector, now $p^\mu(p^2 = 0)$ is a null vector. Therefore let us define another null vector

$$\bar{p}^\mu = (p^0, -\mathbf{p}), \quad p^\mu = (p^0, \mathbf{p}),$$

so that $p^\mu + \bar{p}^\mu$ is a time-like vector, and $p^\mu - \bar{p}^\mu$ is a space-like vector. We now introduce two other unit space-like vectors, $e^\mu{}_{p\lambda}$, which obey

$$e^\mu{}_{p\lambda} {}^* e_{\mu p \lambda'} = \delta_{\lambda\lambda'}, \quad \lambda, \lambda' = 1, 2,$$

$$p_\mu e^\mu{}_{p\lambda} = 0, \quad \bar{p}_\mu e^\mu{}_{p\lambda} = 0. \tag{1.56}$$

Hence

$$g^{\mu\nu} = \frac{(p+\bar{p})^{\mu}(p+\bar{p})^{\nu}}{(p+\bar{p})^{2}} + \frac{(p-\bar{p})^{\mu}(p-\bar{p})^{\nu}}{(p-\bar{p})^{2}} + \sum_{\lambda=1,2} e^{\mu}_{p\lambda} e^{\nu}_{p\lambda}{}^{*},$$

or

$$g^{\mu\nu} = \frac{p^{\mu}\bar{p}^{\nu} + \bar{p}^{\mu}p^{\nu}}{(p\bar{p})} + \sum_{\lambda=1,2} e^{\mu}_{p\lambda} e^{\nu}_{p\lambda}{}^{*}. \tag{1.57}$$

Again the dimensionality can be checked by letting $\mu = \nu$ and summing over μ. One gets

$$4 = 2 + 2.$$

When the representation for $g_{\mu\nu}$, (1.57), is inserted into (1.55), and the conditions $p_{\mu}J^{\mu}(p) = p_{\mu}J^{\mu}(-p) = 0$ are used (notice that $\bar{p}_{\mu}J^{\mu}(p) \neq 0$), we get

$$|\langle 0_{+} | 0_{-}\rangle^{J}|^{2} = \exp\left(-\sum_{p\lambda} |J_{p\lambda}|^{2}\right) \leqslant 1, \tag{1.58}$$

as required. The necessary reduction to two polarizations which are characteristic of the photon has also been accomplished through the use of the divergenceless condition on J^{μ}, (1.52).

1.6   Massless Spin-2 Particles. The Graviton

As a further example to illustrate the universality of the source description, let us discuss briefly another massless boson known (or conjectured) as the graviton. The graviton is a spin-2 particle. Its source is a second-rank tensor obeying

$$T_{\mu\nu}(x) = T_{\nu\mu}(x),$$
$$\partial_\mu T^{\mu\nu}(x) = 0. \tag{1.59}$$

The appropriate expression for the vacuum amplitude is

$$\langle 0_+|0_-\rangle^T = \exp\{(i/2)\int(dx)(dx')T^{\mu\nu}(x)[g_{\mu\lambda}g_{\nu\kappa}$$
$$-(1/2)g_{\mu\nu}g_{\lambda\kappa}]D_+(x-x')T^{\lambda\kappa}(x')\}. \tag{1.60}$$

We must, of course, verify that this expression satisfies the positiveness requirement and at the same time that it involves only the two independent polarizations characteristic of a massless particle.
It follows from (1.60) that

$$|\langle 0_+|0_-\rangle^T|^2 = \exp\{-\int d\omega_p T^{\mu\nu}(-p)[g_{\mu\lambda}g_{\nu\kappa}$$
$$-(1/2)g_{\mu\nu}g_{\lambda\kappa}]T^{\lambda\kappa}(p)\}, \tag{1.61}$$

where $T^{\mu\nu}(p)$ obeys

$$p_\mu T^{\mu\nu}(p) = p_\mu T^{\mu\nu}(-p) = 0.$$

If we insert the dyadic representation for $g_{\mu\nu}$, (1.57) (for simplicity, real e_λ's are used here), we have effectively, for a fixed p,

$$g_{\mu\lambda}g_{\nu\kappa}-(1/2)g_{\mu\nu}g_{\lambda\kappa}\rightarrow\sum_{\alpha,\beta=1,2}\left[e_{\mu\alpha}e_{\nu\beta}e_{\lambda\alpha}e_{\kappa\beta}-(1/2)e_{\mu\alpha}e_{\nu\alpha}e_{\lambda\beta}e_{\kappa\beta}\right].$$

Keeping in mind that symmetry in μ, ν and λ, κ is required, we get finally

$$g_{\mu\lambda}g_{\nu\kappa}-(1/2)g_{\mu\nu}g_{\lambda\kappa}\rightarrow\sum_{\alpha,\beta=1,2}e_{\mu\nu\alpha\beta}e_{\lambda\kappa\alpha\beta}, \tag{1.62}$$

where

$$e_{\mu\nu\alpha\beta} = (1/2)(e_{\mu\alpha}e_{\nu\beta}+e_{\mu\beta}e_{\nu\alpha})-(1/2)\delta_{\alpha\beta}\sum_{\gamma=1,2}e_{\mu\gamma}e_{\nu\gamma}, \tag{1.63}$$

is symmetrical in α and β (also symmetrical in μ and ν), and obeys

$$\sum_{\alpha=1,2}e_{\mu\nu\alpha\alpha} = 0. \tag{1.64}$$

Therefore, only two of the $e_{\mu\nu\alpha\beta}$, $\alpha,\beta=1,2$ are independent, and if one defines

$$\frac{1}{\sqrt{2}}(e_{\mu\nu11}-e_{\mu\nu22}) = e_{\mu\nu1}, \tag{1.65}$$

and

$$\sqrt{2}\, e_{\mu\nu 12} = e_{\mu\nu 2}, \tag{1.66}$$

then

$$\sum_{\alpha,\beta=1,2} e_{\mu\nu\alpha\beta} e_{\lambda\kappa\alpha\beta} = \sum_{\Lambda=1,2} e_{\mu\nu\Lambda} e_{\lambda\kappa\Lambda}. \tag{1.67}$$

With the definition

$$T_{p\Lambda} = \sqrt{d\omega_p}\, e_{\mu\nu p\Lambda} T^{\mu\nu}(p), \tag{1.68}$$

(1.61) becomes finally

$$|\langle 0_+ | 0_- \rangle^T|^2 = \exp\left(-\sum_{p,\Lambda=1,2} |T_{p\Lambda}|^2\right) \leqslant 1, \tag{1.69}$$

which confirms the physical positiveness requirement and exhibits the two independent polarizations characteristic of massless particles.

1.7   Spin-(1/2) Particles

In order to discuss spin-(1/2) particles, we will briefly review the Dirac algebra in the Majorana representation which is particularly useful for our purpose. The 4×4 matrices γ^μ obey

$$\{\gamma^\mu, \gamma^\nu\} = -2g^{\mu\nu}, \tag{1.70}$$

$$(\gamma^0)^2 = +1, \ (\gamma_k)^2 = -1, \ k = 1, 2, 3. \tag{1.71}$$

These matrices possess simple algebraic properties. But as a set they do not have a common symmetry, γ^0 is imaginary and antisymmetrical (and therefore Hermitian), while γ_k ($k = 1, 2, 3$) is imaginary and symmetrical (and therefore skew Hermitian). However, out of these matrices one can construct another set which do have a common symmetry. But their algebraic properties are no longer simple. These are $\gamma^0\gamma^\mu$ which are all real and symmetrical. The fact that one cannot have a set of matrices with both simple algebraic properties and a common symmetry is due to the indefinite nature of the space-time metric. The matrix γ^0 represents the fundamental metric tensor. In the same sense that $J^\mu g_{\mu\nu} J^\nu$ is a scalar, so is $\eta\gamma^0\eta$ if η is a four-component spinor.

As I have remarked before, the basic exponential quadratic form of $\langle 0_+|0_-\rangle$ is entirely general. Thus we ask: Is it correct to simply write

$$\langle 0_+|0_-\rangle^\eta = \exp\left[(i/2)\int(dx)(dx')\eta(x)\gamma^0\Delta_+(x-x')\eta(x')\right]$$

in order to describe spin-(1/2) particles? The objection here is that all four components of $\eta_\zeta(x)$ contribute, giving a multiplicity of four rather than the multiplicity of two appropriate to spin-(1/2). This is overcome by writing

$$\langle 0_+|0_-\rangle^\eta = \exp\left[(i/2)\int(dx)(dx')\eta(x)\gamma^0(m - \gamma^\mu(1/i)\partial_\mu)\right.$$
$$\left. \times \Delta_+(x-x')\eta(x')\right]. \tag{1.72}$$

As usual, consider an emission and an absorption source,

$$\eta = \eta_1 + \eta_2.$$

Then

$$i\int(dx)(dx')\eta_1(x)\gamma^0(m - \gamma(1/i)\partial)\Delta_+(x-x')\eta_2(x')$$
$$= \int d\omega_p i\eta_1(-p)\gamma^0(m - \gamma p)\eta_2(p). \tag{1.73}$$

For a particular momentum p, this structure can be examined in its rest frame ($\mathbf{p} = 0$, $p^0 = m$), and

$$m - \gamma p \to m + \gamma^0 m = m(1 + \gamma^0), \tag{1.74}$$

which selects only $\gamma^{0'} = +1$, and rejects $\gamma^{0'} = -1$. This produces the required reduction to two independent sources, characteristic of spin-(1/2) particles, since $\gamma^{0'} = +1$ is a property unaltered by spatial rotations.

We now define the 4×4 matrix function

$$G_+(x-x') = (m - \gamma^\mu(1/i)\partial_\mu)\Delta_+(x-x'), \tag{1.75}$$

and observe that

$$(\gamma^\mu(1/i)\partial_\mu + m)G_+(x-x') = \delta(x-x'), \tag{1.76}$$

in virtue of the differential equation satisfied by $\Delta_+(x-x')$. Thus, $G_+(x-x')$ is the Green's function of the Dirac equation. Transposition of the matrices, combined with the exchange of x and x', has the following effect on $\gamma^0 G_+(x-x')$,

$$[\gamma^0 G_+(x'-x)]^T = (\gamma^0 m - \gamma^0 \gamma^\mu(1/i)\partial'_\mu)^T \Delta_+(x'-x)$$

$$= -[\gamma^0 G_+(x-x')], \tag{1.77}$$

since γ^0 is antisymmetrical and the $\gamma^0 \gamma^\mu$ are symmetrical ($\partial'_\mu = -\partial_\mu$). But then if $\eta(x)$ are ordinary numbers, the quadratic form in the exponential function of (1.72) is zero identically! There are two possibilities to escape this conclusion. The first one is to introduce another matrix in $G_+(x-x')$, which is outside of space-time and therefore does not affect the spin description, and is antisymmetrical. The simplest choice is

$$q = \begin{pmatrix} 0 & -i \\ i & 0 \end{pmatrix}. \tag{1.78}$$

The source functions, $\eta(x)$, of course are now given an additional multiplicity of two. Keeping the previous definition of $G_+(x-x')$, we then have

$$\langle 0_+ | 0_- \rangle^\eta = \exp[(i/2)\int (dx)(dx')\eta(x)\gamma^0 q G_+(x-x')\eta(x')], \tag{1.79}$$

which describes spin-(1/2) BE particles. But the physical requirement on the vacuum persistence probability that it should never exceed unity, which is a positiveness requirement,

$$\int (dx)(dx')\eta(x)\gamma^0 q \, \mathrm{Re} \, (1/i)G_+(x-x')\eta(x') \geqslant 0 \tag{1.80}$$

can never be satisfied! The matrix q, being intrinsically indefinite, has eigenvalues ± 1, and is independent of space-time matrices. Therefore a matrix transformation on q can change q into $-q$, and the left-hand side of (1.80) will change sign, in contradiction with the positiveness requirement.

The second possibility is to match the multiplication property of the sources $\eta_\zeta(x)$ to the *antisymmetry* of the structure $\gamma^0 G_+(x-x')$, i.e. they must be anticommuting numbers:

$$\eta_\zeta(x)\eta_{\zeta'}(x') = -\eta_{\zeta'}(x')\eta_\zeta(x).$$

Now we have an identity, instead of a paradox,

$$\int (dx)(dx')\sum_{\zeta,\zeta'}\eta_\zeta(x)[\gamma^0 G_+(x-x')]_{\zeta\zeta'}\eta_{\zeta'}(x')$$
$$= -\int (dx)(dx')\sum_{\zeta,\zeta'}\eta_\zeta(x)[\gamma^0 G_+(x'-x)]_{\zeta'\zeta}\eta_{\zeta'}(x')$$
$$= +\int (dx)(dx')\sum_{\zeta,\zeta'}\eta_{\zeta'}(x')[\gamma^0 G_+(x'-x)]_{\zeta'\zeta}\eta_\zeta(x),$$

which is simply a relabelling of variables. Just this kind of anticommuting number is well-known as the elements of a Grassman, or exterior algebra.

The physical properties of the system can now be studied by the standard procedure. Consider an emission and absorption source

$$\eta = \eta_1 + \eta_2,$$

and deduce

$$\langle 0_+|0_-\rangle^\eta = \langle 0_+|0_-\rangle^{\eta_1}$$
$$\times \exp[i\int (dx)(dx')\eta_1(x)\gamma^0 G_+(x-x')\eta_2(x')]\langle 0_+|0_-\rangle^{\eta_2}, \quad (1.81)$$

since quadratic functions of the anticommuting numbers are commutative. Now

$$x^0 > x^{0'}: G_+(x-x') = i\int d\omega_p e^{ip(x-x')}(m-\gamma p), \quad (1.82)$$

where the projection matrix has the property

$$\left(\frac{m-\gamma p}{2m}\right)^2 = \frac{m-\gamma p}{2m}, \quad (1.83)$$

since

$$(\gamma p)^2 = m^2. \quad (1.84)$$

The projection matrix can be exhibited in the eigenfunction expansion

$$\frac{m-\gamma p}{2m} = \sum_{\lambda=1,2} u_{p\lambda} u_{p\lambda}^* \gamma^0, \quad (1.85)$$

where

$$u_{p\lambda}^* \gamma^0 u_{p\lambda'} = \delta_{\lambda\lambda'}, \quad (1.86)$$

and λ is a polarization or spin index. The dimensionality can be checked by taking the trace of (1.85).

$$2 = \sum_{\lambda=1,2} u_{p\lambda}^* \gamma^0 u_{p\lambda} = 2, \tag{1.87}$$

in which the relation $\mathrm{Tr}\, \gamma_\mu = 0$ has been used. Thus,

$$x^0 > x^{0\prime}: G_+(x-x') = i\int d\omega_p e^{ip(x-x')} 2m \sum_{\lambda=1,2} u_{p\lambda} u_{p\lambda}^* \gamma^0. \tag{1.88}$$

We now define

$$\sqrt{d\omega_p 2m}\, u_{p\lambda}^* \gamma^0 \eta(p) = \eta_{p\lambda}, \tag{1.89}$$

$$\sqrt{d\omega_p 2m}\, \eta(-p)\gamma^0 u_{p\lambda} = \eta_{p\lambda}^*.$$

It is taken for granted that we keep the convention of choosing $\eta(x)$ to be real. Later we will learn more about the complex conjugation property of the anticommuting numbers. As linear combinations of the $\eta(x)$, $\eta_{p\lambda}$ and $\eta_{p\lambda}^*$ are totally anticommutative, i.e.,

$$\{\eta_{p\lambda}, \eta_{p'\lambda'}\} = 0, \tag{1.90}$$

which implies, in particular,

$$(\eta_{p\lambda})^2 = 0. \tag{1.91}$$

Hence

$$\exp\left[i\int (dx)(dx')\eta_1(x)\gamma^0 G_+(x-x')\eta_2(x')\right]$$

$$= \exp\left(\sum_{p\lambda} i\eta_{1p\lambda}^* i\eta_{2p\lambda}\right)$$

$$= \prod_{p\lambda} \exp\left(i\eta_{1p\lambda}^* i\eta_{2p\lambda}\right)$$

$$= \prod_{p\lambda} \left(1 + i\eta_{1p\lambda}^* i\eta_{2p\lambda}\right). \tag{1.92}$$

That is

$$\exp\left(\sum_{p\lambda} i\eta_{1p\lambda}^* i\eta_{2p\lambda}\right) = \prod_{p\lambda} \sum_{n_{p\lambda}} \left(i\eta_{1p\lambda}^* i\eta_{2p\lambda}\right)^{n_{p\lambda}}, \tag{1.93}$$

where each $n_{p\lambda}$ can take only on two values,

$$n_{p\lambda} = 0, 1.$$

We recognize the basic feature of Fermi-Dirac (FD) statistics, the exclusion principle! A FD source cannot operate to create two particles with the same properties. The vacuum amplitude has the general form

$$\langle 0_+ | 0_- \rangle^\eta = \langle 0_+ | 0_- \rangle^{\eta_1} \prod_{p\lambda} \sum_{n_{p\lambda}=0,1} \left(i\eta_{1p\lambda}^* i\eta_{2p\lambda}\right)^{n_{p\lambda}} \langle 0_+ | 0_- \rangle^{\eta_2}. \tag{1.94}$$

Since the sources do have simple algebraic properties, each term of this structure can be factored into two parts, one on the left referring entirely to the detection source and the other on the right referring entirely to the emission source. To illustrate the procedure, consider the example of two states, a and b,

$$i\eta_{1a}{}^* i\eta_{2a} i\eta_{1b}{}^* i\eta_{2b}$$

$$= (i\eta_{1b}{}^* i\eta_{1a}{}^*)(i\eta_{2a} i\eta_{2b}),$$

and in general

$$\langle \{n\}|0_-\rangle^\eta = \prod_{p\lambda} \frac{(i\eta_{p\lambda})^{n_{p\lambda}}}{\sqrt{n_{p\lambda}!}} \langle 0_+|0_-\rangle^\eta,$$

$$\langle 0_+|\{n\}\rangle^\eta = \langle 0_+|0_-\rangle^\eta \prod_{p\lambda}{}^T \frac{(i\eta_{p\lambda}{}^*)^{n_{p\lambda}}}{\sqrt{n_{p\lambda}!}}, \tag{1.95}$$

where $\prod$ is some standard multiplication order and $\prod^T$ is its inverse. The factor $n_{p\lambda}! = 1$ is not necessary, it is included only to show that the results are universal—BE particles and FD particles are exactly on the same footing, only the implicit algebraic property of the sources conveying the nature of the particles.

It is important to check the completeness property of the multiparticle states, as in

$$\sum_n \langle 0_-|n\rangle^\eta \langle n|0_-\rangle^\eta = \sum_n [\langle n|0_-\rangle^\eta]^* \langle n|0_-\rangle^\eta = 1,$$

$$\sum_n \langle 0_+|n\rangle^\eta \langle n|0_+\rangle^\eta = \sum_n \langle 0_+|n\rangle^\eta [\langle 0_+|n\rangle^\eta]^* = 1, \tag{1.96}$$

where n symbolizes the whole set of occupation numbers. The sense of multiplication here is important, since the individual factors are functions of noncommuting numbers. On inserting the expressions (1.95), these conditions become

$$1 = |\langle 0_+|0_-\rangle^\eta|^2 \sum_n \left[\prod_{p\lambda}(\eta_{p\lambda})^{n_{p\lambda}}\right]^* \prod_{p\lambda}(\eta_{p\lambda})^{n_{p\lambda}},$$

$$1 = |\langle 0_+|0_-\rangle^\eta|^2 \sum_n \prod_{p\lambda}{}^T (\eta_{p\lambda}{}^*)^{n_{p\lambda}} \left[\prod_{p\lambda}{}^T (\eta_{p\lambda}{}^*)^{n_{p\lambda}}\right]^*. \tag{1.97}$$

These two relations become identical, if complex conjugation reverses the sense of multiplication, as conveyed by

$$(\eta_\zeta(x)\eta_{\zeta'}(x'))^* = \eta_{\zeta'}(x')\eta_\zeta(x). \tag{1.98}$$

This is a general source property, since transposition is ineffective for BE sources. We infer from (1.98) that, while the product of two real BE sources is real, the product of two real FD sources is imaginary. Now

$$\left[\prod_{p\lambda}(\eta_{p\lambda})^{n_{p\lambda}}\right]^* \prod_{p\lambda}(\eta_{p\lambda})^{n_{p\lambda}}$$

$$= \prod_{p\lambda}^{T}(\eta_{p\lambda}{}^*)^{n_{p\lambda}}\prod_{p\lambda}(\eta_{p\lambda})^{n_{p\lambda}}$$

$$= \prod_{p\lambda}(\eta_{p\lambda}{}^*\eta_{p\lambda})^{n_{p\lambda}},$$

and (1.97) becomes

$$1 = |\langle 0_+|0_-\rangle^{\eta}|^2 \exp(\sum_{p\lambda}\eta_{p\lambda}{}^*\eta_{p\lambda}), \tag{1.99}$$

which is thus valid for both statistics. Now, direct evaluation gives

$$|\langle 0_+|0_-\rangle^{\eta}|^2 = \exp\Big[- \text{Re} \int(dx)(dx')\eta(x)\gamma^0(m-\gamma(1/i)\partial)$$
$$\times (1/i)\varDelta_+(x-x')\eta(x')\Big].$$

From what we have just learned, the combination $\eta(x)\gamma^0\eta(x')$ is real since γ^0 is imaginary. Also $(1/i)\gamma_\mu$ is real. Therefore, everything is real except $(1/i)\varDelta_+(x-x')$. But the real part of the latter is

$$\text{Re}\,(1/i)\varDelta_+(x-x') = \text{Re}\int d\omega_p e^{ip(x-x')}.$$

Under the integration sign, we have

$$m-\gamma(1/i)\partial\rightarrow m-\gamma p = 2m\sum_\lambda u_{p\lambda}u_{p\lambda}{}^*\gamma^0.$$

Therefore,

$$|\langle 0_+|0_-\rangle^{\eta}|^2 = \exp(- \text{Re}\sum_{p\lambda}\eta_{p\lambda}{}^*\eta_{p\lambda}). \tag{1.100}$$

But

$$(\eta_{p\lambda}{}^*\eta_{p\lambda})^* = \eta_{p\lambda}{}^*\eta_{p\lambda}, \tag{1.101}$$

is a reality statement, and the completeness requirement is verified.

Spin (1/2) is only compatible with FD statistics. We now turn to the question of compatibility between integer spin and FD statistics. The actual integer spin value is irrelevant in this consideration, and the essence of the whole argument can be seen from the simplest case, spin 0. First we notice that we cannot let the source functions, $K(x)$, be anticommuting numbers, without introducing an antisymmetrical matrix q to compensate the fundamental symmetry of the function $\varDelta_+(x-x')$ in x and x'. This matrix q must be a non-space-time one, in order not to affect the already established spin description. We give $K(x)$ a multiplicity of two, and the matrix q is that introduced previously. Thus, spin-0 FD particles are described by

$$\langle 0_+ | 0_- \rangle^K = \exp\left[(i/2) \int (dx)(dx') K(x) q \Delta_+(x-x') K(x') \right]. \quad (1.102)$$

Write

$$q = \sum_{q'=\pm 1} q' u_{q'} \cdot u_{q'}{}^*, \quad (1.103)$$

and define

$$u_{q'}{}^* \sqrt{d\omega_p} K(p) = K_{pq'},$$

$$\sqrt{d\omega_p} K(-p) u_{q'} = K_{pq'}{}^*, \quad (1.104)$$

for the consideration of an emission and absorption source. Accordingly

$$i \int (dx)(dx') K_1(x) q \Delta_+(x-x') K_2(x')$$
$$= \sum_{pq'} i K_{1pq'}{}^* q' i K_{2pq'}.$$

But

$$q' = \left[e^{i \left[\frac{(1-q')}{4} \right] \pi} \right]^2, \quad (1.105)$$

only introduces a q'-dependent phase factor in the identification of multi-particle states. These phase factors will be suppressed when we consider the completeness requirement (1.29), since each matrix element is combined with its complex conjugate. So it is required that

$$1 = |\langle 0_+ | 0_- \rangle^K|^2 \exp(\sum_{pq'} K_{pq'}{}^* K_{pq'}). \quad (1.106)$$

But since $K(x) q K(x')$ is real, which follows from (1.98), we have directly

$$|\langle 0_+ | 0_- \rangle^K|^2 = \exp\left[- \int (dx)(dx') K(x) q \, \mathrm{Re}\, (1/i) \Delta_+(x-x') K(x') \right]$$
$$= \exp(-\sum_{pq'} q' K_{pq'}{}^* K_{pq'}), \quad (1.107)$$

in which q' appears explicitly. The completeness requirement is violated through the indefiniteness of q.

1.8 Multispinor Formulation of Arbitrary Spin. General Connection of Spin and Statistics

We have seen in the specific examples of spin-0 and spin-(1/2) that there is a definite connection between spin and statistics. We will now show that the connection of spin and statistics can be established for arbitrary spin. Indeed, there is a uniform treatment of all spins, which gives the spin-statistics connection in a general way. It uses multispinors $S_{\zeta_1 \ldots \zeta_n}(x)$, each $\zeta = 1, 2, 3, 4$ being a Dirac index. Using the spin-(1/2) structure repeated n times, one gets

$$\langle 0_+ | 0_- \rangle^S = \exp\left[(i/2) \int (dx)(dx') S(x) \prod_{\alpha=1}^{n} \gamma_\alpha^0 \prod_{\alpha=1}^{n} \right.$$
$$\left. \times (m - \gamma^\mu (1/i)\partial_\mu)_\alpha \Delta_+(x-x') S(x') \right], \tag{1.108}$$

where

$$[\gamma_\alpha^\mu, \gamma_\beta^\nu] = 0, \text{ if } \alpha \neq \beta, \tag{1.109}$$

since they act on different indices. The consideration of an emission and absorption source

$$S = S_1 + S_2$$

gives

$$\int (dx)(dx') S_1(x) \prod_\alpha \left[\gamma^0 (m - \gamma^\mu (1/i)\partial_\mu) \right]_\alpha \Delta_+(x-x') S_2(x')$$
$$= \int d\omega_p i S_1(p)^* \prod_\alpha \left[\gamma^0 (m - \gamma p) \right]_\alpha i S_2(p). \tag{1.110}$$

For a specified momentum, in its rest frame we have

$$\prod_\alpha \left[\gamma^0 (m - \gamma p) \right]_\alpha = \prod_\alpha \left[\gamma^0 (m + \gamma^0 m) \right]_\alpha = \prod_\alpha \left[m(1 + \gamma^0) \right]_\alpha, \tag{1.111}$$

which is the projection matrix to select $\gamma^{0'} = +1$ for each index α. Thus, in the rest frame, the sources reduce effectively to $S_{\sigma_1 \sigma_2 \ldots \sigma_n}$, each σ being a two-valued spin index. Now suppose $S_{\zeta_1 \zeta_2 \ldots \zeta_n}$ has definite symmetry property. Specifically, let us consider the totally symmetrical one. Then $S_{\sigma_1 \sigma_2 \ldots \sigma_n}$ is totally symmetrical. As we all have learned from elementary quantum mechanics, this produces the state of maximum resultant spin, that is, the sources are those of a particle of spin

$$s = (1/2)n, \tag{1.112}$$

and

$$s = (1/2), 1, (3/2), \ldots,$$

as

$$n = 1, 2, 3, \ldots.$$

Only $s = 0$ is missing in this series. For this one has to take $n = 2$ and use an antisymmetrical spinor source $S_{\zeta_1\zeta_2}(x)$. Incidentally, for $n \geqslant 3$, the independent components of a totally antisymmetrical multispinor vanishes identically in the rest frame, since $S_{\sigma_1\sigma_2\ldots\sigma_n} = 0 (n \geqslant 3)$. Note that more complicated symmetry patterns give equivalent descriptions, to the extent that a definite spin appears in the rest frame. Consider, for example, $n = 3$ with the requirement of antisymmetry in a pair of Dirac indices. The latter contributes zero spin in the rest frame and we have a possible description of an $s = (1/2)$ particle.

The symmetry property of the matrix propagation function is expressed by

$$\left\{\prod_\alpha \left[\gamma^0(m - \gamma^\mu(1/i)\partial'_\mu)\right]_\alpha \Delta_+(x' - x)\right\}^T$$
$$= \prod_\alpha (-)\left[\gamma^0(m - \gamma^\mu(1/i)\partial_\mu)\right]_\alpha \Delta_+(x - x'),$$

or

$$\left\{\prod_\alpha \left[\gamma^0(m - \gamma^\mu(1/i)\partial'_\mu)\right]_\alpha \Delta_+(x' - x)\right\}^T$$
$$= (-1)^n \prod_\alpha \left[\gamma^0(m - \gamma^\mu(1/i)\partial_\mu)\right]_\alpha \Delta_+(x - x'). \qquad (1.113)$$

If the algebraic properties of the sources are to match the symmetry properties of the propagation function, we learn that

$$n = \text{even}, \; s = \text{integer}: \left[S(x), S(x')\right] = 0, \; \text{BE}$$

$$n = \text{odd}, \; s = \text{integer} + (1/2): \left\{S(x), S(x')\right\} = 0, \; \text{FD} \qquad (1.114)$$

which express the general connection between spin and statistics.

In the same way as for the case spin-$(1/2)$, one can verify that the completeness requirement

$$\sum_{\{n\}} \langle 0_- | \{n\} \rangle^s \langle \{n\} | 0_- \rangle^s = 1, \qquad (1.115)$$

is satisfied, which is shown by alternatively using $\langle 0_+ | 0_- \rangle^s$ to calculate directly $|\langle 0_+ | 0_- \rangle^s|^2$ and indirectly by producing first $\langle \{n\} | 0_- \rangle^s / \langle 0_+ | 0_- \rangle^s$. The consistency test of the two procedures is important for attempts to reverse the natural spin-statistics connection by introducing a non-space-time antisymmetrical matrix $q = \begin{pmatrix} 0 & -i \\ i & 0 \end{pmatrix}$. The completeness condition can be rewritten as

$$1 = \sum_{\{n\}} \frac{\langle 0_- | \{n\} \rangle^s}{\langle 0_- | 0_+ \rangle^s} \cdot \frac{\langle \{n\} | 0_- \rangle^s}{\langle 0_+ | 0_- \rangle^s} |\langle 0_+ | 0_- \rangle^s|^2. \qquad (1.116)$$

As in the discussion of the previous section for spin 0, eigenvalues q' appear as phase factors in the identification of multiparticle states, since $q' = 1$, or $q' = -1 = (i)^2$. These phase factors disappear on forming

$$\frac{\langle 0_- | \{n\} \rangle^S}{\langle 0_- | 0_+ \rangle^S} \cdot \frac{\langle \{n\} | 0_- \rangle^S}{\langle 0_+ | 0_- \rangle^S} = \left[\frac{\langle \{n\} | 0_- \rangle^S}{\langle 0_+ | 0_- \rangle^S} \right]^*$$

$$\times \left[\frac{\langle \{n\} | 0_- \rangle^S}{\langle 0_+ | 0_- \rangle^S} \right]. \qquad (1.117)$$

The direct evaluation of $|\langle 0_+ | 0_- \rangle^S|^2$ gives

$$|\langle 0_+ | 0_- \rangle^S|^2 = \exp\{ - \int (dx)(dx') S(x) q \prod_\alpha [\gamma^0(m - \gamma(1/i)\partial)]_\alpha$$

$$\times \text{Re} \,(1/i) \Delta_+(x - x') S(x')\}, \qquad (1.118)$$

since $S(x) q \prod_\alpha \gamma^0{}_\alpha S(x')$ is real with abnormal statistics, while $S(x) \prod_\alpha \gamma^0{}_\alpha S(x')$ is real with normal statistics. In both cases, the reality property follows simply from the Hermitian character of the matrices. Now the indefinite eigenvalues of q are explicitly involved in (1.118), and the completeness requirement is violated.

2. ELECTRODYNAMICS

2.1 Combined System of Noninteracting Electrons and Photons. Fields

We shall now study the most familiar dynamical system, of electrons and photons, in order to indicate how the interaction among particles can be incorporated very naturally into the source theory.

Properties of noninteracting particles are described by individual vacuum amplitudes. They are multiplied together to describe the combined system. Write

$$\langle 0_+|0_-\rangle^S = e^{iw(S)}, \tag{2.1}$$

where, for the system of electrons and photons under physical conditions of noninteraction,

$$w(S) = w_2 = (1/2)\int (dx)(dx')\eta(x)\gamma^0 G_+(x-x')\eta(x')$$
$$+ (1/2)\int (dx)(dx')J^\mu(x)D_+(x-x')J_\mu(x'). \tag{2.2}$$

It is convenient to introduce auxiliary quantities which express the effects at one source of other sources, and combine the latter with the particular propagation function. Thus, we define

$$\psi(x) = \int (dx')G_+(x-x')\eta(x'), \; (\gamma^\mu(1/i)\partial_\mu+m)\psi(x) = \eta(x),$$
$$A_\mu(x) = \int (dx')D_+(x-x')J_\mu(x'), -\partial^2 A_\mu(x)$$
$$= J_\mu(x), \; \partial_\mu A^\mu(x) = 0. \tag{2.3}$$

29

These are the fields of the corresponding sources. They are numerical quantities of the same type as the sources, namely, $A_\mu(x)$ are totally commutative numbers, and $\psi(x)$ are totally anticommutative. The auxiliary quantities, fields, are very useful, since they summarize the effects, in the region of interest, of sources which may be very far away. Now, w_2 can be represented alternatively as

$$w_2 = (1/2)\int (dx)\eta(x)\gamma^0\psi(x) + (1/2)\int (dx)J^\mu(x)A_\mu(x)$$

$$= (1/2)\int (dx)\psi(x)\gamma^0\eta(x) + (1/2)\int (dx)J^\mu(x)A_\mu(x), \qquad (2.4)$$

or

$$w_2 = (1/2)\int (dx)\psi(x)\gamma^0(\gamma^\mu(1/i)\partial_\mu + m)\psi(x)$$

$$+ (1/2)\int (dx)A^\mu(x)(-\partial^2)A_\mu(x). \qquad (2.5)$$

In the latter form only fields appear. A third form combines the two representations,

$$w_2 = \int (dx)\left[\psi(x)\gamma^0\eta(x) - (1/2)\psi(x)\gamma^0(\gamma(1/i)\partial + m)\psi(x)\right]$$

$$+ \int (dx)\left[A^\mu(x)J_\mu(x) - (1/2)A^\mu(x)(-\partial^2)A_\mu(x)\right]. \qquad (2.6)$$

All the three forms are equivalent in virtue of the relations between fields and sources, (2.3). But the last form has the advantage that these connections can be derived from it instead of being stated independently. Let us examine the dependence of w_2 on the sources η, J_μ and the fields ψ, A_μ,

$$\delta w_2 = \int (dx)\{\psi(x)\gamma^0\delta\eta(x) + \delta\psi(x)\gamma^0\left[\eta(x) - (\gamma(1/i)\partial + m)\psi(x)\right]$$

$$+ A^\mu(x)\delta J_\mu(x) + \delta A^\mu(x)\left[J_\mu(x) - (-\partial^2)A_\mu(x)\right]\}.$$

Now, if we insist that w_2 depends only on the sources, that is, that it be stationary with respect to variations of the fields, the field differential equations are recovered. In this form w_2 is an action, and we write

$$w_2 = \int (dx)\left[\psi(x)\gamma^0\eta(x) + A^\mu(x)J_\mu(x) + \mathcal{L}(\psi, A)\right], \qquad (2.7)$$

where $\mathcal{L}$ is the Lagrange function of the system,

$$\mathcal{L} = -(1/2)\psi\gamma^0(\gamma(1/i)\partial + m)\psi - (1/2)A^\mu(-\partial^2)A_\mu. \qquad (2.8)$$

Since $\partial_\mu A^\mu(x) = 0$, the second term can also be replaced by

$$-(1/2)\left[A^\mu(-\partial^2)A_\mu - (\partial_\mu A^\mu)^2\right]$$

$$\to -(1/2)\left[(\partial_\mu A_\nu)^2 - (\partial_\mu A^\mu)^2\right]$$

$$\to -(1/4)F^{\mu\nu}F_{\mu\nu}, \qquad (2.9)$$

as far as w_2 is concerned, where

$$F_{\mu\nu} = \partial_\mu A_\nu - \partial_\nu A_\mu. \tag{2.10}$$

The Lagrange function thus acquires the new form

$$\mathscr{L} = -(1/2)\psi\gamma^0(\gamma(1/i)\partial + m)\psi - (1/4)F^{\mu\nu}F_{\mu\nu}. \tag{2.11}$$

Since the source J_μ is conserved

$$\partial_\mu J^\mu(x) = 0,$$

we now have the freedom of gauge transformation

$$A_\mu(x) \rightarrow A_\mu(x) + \partial_\mu \lambda(x), \ \lambda(x) \text{ arbitrary,}$$

which leaves the action w_2 unchanged. Gauge invariance is an important aspect of the masslessness of the photon.

2.2   Primitive Interaction and Interaction Skeleton. Extended Sources.

So far sources are only called upon to liberate energy-momentum, or mass, equal to that of the particle being described. This is too restrictive for a dynamical theory. The notion of source is an abstraction of realistic physical production and annihilation mechanisms. The simplest example is provided by the creation of a charged particle in a collision, which involves the transfer of charge to the particle from the source (idealizing all the partners in the collision). An accelerated charge radiates, and the possible creation of a photon along with the particle is an inescapable part of the physical mechanism. Thus it would be artificial to separate the mechanism that creates a charged particle alone from the mechanism that creates a charged particle and a photon. Accordingly we regard it meaningful and useful to define a charged particle source $\eta(p)$, not only for $-p^2 = m^2$, but also for $-p^2 > m^2$. Thus, we now have two possible emission processes for a charged particle source, as shown in Fig. 2.1, where the emission of an

FIGURE 2.1

$$-p^2 = m^2 \qquad\qquad -p^2 > m^2$$

electron and a photon is supposed to occur without interaction between these particles. We shall call the latter the primitive interaction. A similar discussion also applies to an absorption source. These considerations suggest more generally that it is meaningful and useful to generalize a source to include any combination of particles with the same properties, apart from mass, as the specified particle. This is the general concept of an extended source.

The primitive interaction, describing the dynamical process in which an electron source $\eta(p)$, with $-p^2 > m^2$, radiates an electron and a photon, gives new possibilities for simple particle sources. For example, the subsequent removal of the photon, or of the electron, gives a particular realization of an electron source, or of a photon source, as illustrated in Fig. 2.2. Single particle exchanges between these sources give new phenomena such as the ones shown in Fig. 2.3. These couplings involve four sources, $\eta\eta JJ$, or $\eta\eta\eta\eta$, which can be written in space-time forms in a way that permits immediate generalization to a wider class of spatio-temporal arrangements. In particular, the final forms can be used to give a first, skeleton description of electron-photon scattering or electron (positron)-electron scattering

FIGURE 2.2

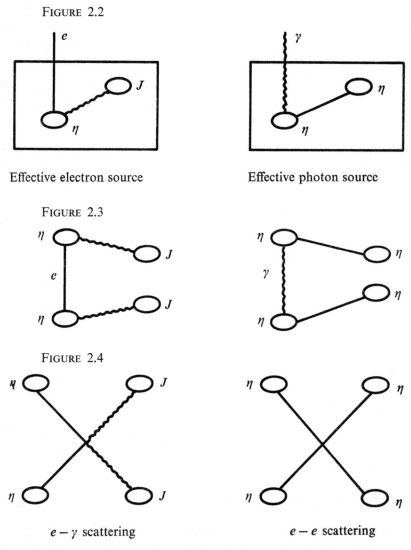

Effective electron source   Effective photon source

FIGURE 2.3

FIGURE 2.4

$e - \gamma$ scattering $e - e$ scattering

(Fig. 2.4). This process can be continued. For example, from electron-photon scattering, one gets an effective electron source of the form ηJJ. Single electron exchange with an effective electron source of the form ηJ gives a new process which involves $\eta\eta JJJ$, and so on. Thus, one gets a series of dynamical processes involving $\eta\eta J, \eta\eta JJ, \eta\eta JJJ, \ldots$ etc. This proliferation introduces, at a first stage, an infinite variety of interactions, in skeletal form, involving only single particle exchanges.

One can give an analytic expression to describe all these processes by beginning with the primitive interaction term w_3 which must have the form

$$w_3 = (1/2)\int (dx)(dx')(d\xi)\eta(x)\gamma^0 G(xx', \xi)^\mu \eta(x')A_\mu(\xi), \qquad (2.12)$$

from the consideration of an effective photon source. But we must also be able to use the effective electron source point of view. Therefore each η must be associated with a G_+ that describes the propagation of the electron to the point of detection. Thus, w_3 must have the form

$$w_3 \sim \int (dx)\psi(x)\psi(x)A(x)$$

omitting the finer details. Now, independently, suppose we do have an additional term in w, say w', that depends on A_μ. Let us change J_μ, and therefore A_μ, by an infinitesimal function. Thus

$$w = w_2 + w'(A), \qquad (2.13)$$

and

$$\delta w'(A) = \int (d\xi)(d\xi')\delta J^\mu(\xi)D_+(\xi-\xi')\frac{\delta w'(A)}{\delta A_\mu(\xi')}, \qquad (2.14)$$

which can be interpreted as single photon exchange between $\delta J^\mu(\xi)$ and an effective photon source $\dfrac{\delta w'(A)}{\delta A_\mu(\xi')}$. The latter, as a photon source, must be conserved,

$$\partial_\mu \frac{\delta w'(A)}{\delta A_\mu(x)} = 0, \qquad (2.15)$$

or

$$\int (dx)\lambda(x)\partial_\mu \frac{\delta w'}{\delta A_\mu(x)} = 0, \qquad (2.16)$$

where $\lambda(x)$ is an arbitrary function. Therefore

$$\int (dx)\partial_\mu\lambda(x)\frac{\delta w'}{\delta A_\mu(x)} = 0, \qquad (2.17)$$

which asserts that $w'(A)$ must remain unchanged when $A_\mu(x)$ is displaced by $\partial_\mu\lambda(x)$. In other words, $w'(A)$ must be gauge invariant.

An elementary solution to both requirements is perfectly familiar. It is obtained from w_2 by the gauge-covariant substitution

$$\partial_\mu \to \partial_\mu - ieqA_\mu$$

and has the structure

$$\mathcal{L} = -(1/4)F^{\mu\nu}F_{\mu\nu} - (1/2)\psi\gamma^0\big[\gamma^\mu((1/i)\partial_\mu - eqA_\mu) + m\big]\psi. \qquad (2.18)$$

Here $q = \begin{pmatrix} 0 & -i \\ i & 0 \end{pmatrix}$ is the charge matrix which must be introduced, since $\psi\gamma^0\gamma^\mu\psi \equiv 0$. The symmetry of $\gamma^0\gamma^\mu$ is in conflict with the antisymmetry of $\psi_\zeta\psi_{\zeta'}$. Thus, in order to represent the charge of the electron, it is necessary to introduce an additional multiplicity of two, and ψ is now an eight-component object. The coupling constant e is the observed charge of the electron, and may be identified through soft photon emission. The action w is invariant under the gauge transformation

$$A_\mu \rightarrow A_\mu + \partial_\mu\lambda, \qquad (2.19)$$

$$\psi \rightarrow e^{ieq\lambda}\psi. \qquad (2.20)$$

It is necessary that $\eta(x)$ responds like $\psi(x)$ to a gauge transformation. This expresses the fact that $\eta(x)$ symbolizes the charged particles involved in creating the electron, which also emit and absorb photons.

The logical progression in this development should be noted. We start with the photon. Its masslessness demands that the source be a conserved vector, from which follows the requirement of gauge invariance, which is not an independent hypothesis. In quantum field theory the procedure is exactly reversed. One begins with the requirement of gauge invariance and deduces the existence of the photon. But this also involves a dynamical restriction. If the coupling is sufficiently strong, there is the possibility that a massless particle does not exist.

The action principle gives the field equation for ψ,

$$\big[\gamma((1/i)\partial - eqA) + m\big]\psi = \eta, \qquad (2.21)$$

or

$$(\gamma(1/i)\partial + m)\psi = \eta + \gamma eqA\psi, \qquad (2.22)$$

which can be converted into the integral equation,

$$\psi = G_+\eta + G_+\gamma eqA\psi, \qquad (2.23)$$

in which space-time coordinates are regarded as matrix indices. We shall use the notation ψ_A to distinguish the electron field influenced by the vector potential A from the noninteracting field ψ,

$$\psi_A = \psi + G_+\gamma eqA\psi_A. \qquad (2.24)$$

The iterated solution is

$$\psi_A = \psi + G_+eq\gamma A\psi + G_+eq\gamma A G_+eq\gamma A\psi + \dots, \qquad (2.25)$$

and

$$w = \int (dx) \left[J_\mu A^\mu - (1/4) F^{\mu\nu} F_{\mu\nu} + (1/2) \eta \gamma^0 \psi_A \right]$$
$$= \int (dx) \left[J_\mu A^\mu - (1/4) F^{\mu\nu} F_{\mu\nu} \right]$$
$$+ (1/2) \int (dx) \eta \gamma^0 \psi + (1/2) \int \psi \gamma^0 eq\gamma A \psi$$
$$+ (1/2) \int \psi \gamma^0 eq\gamma A G_+ eq\gamma A \psi + \dots. \tag{2.26}$$

The action principle also supplies the field equation for A (in the Lorentz gauge, $\partial_\mu A^\mu = 0$),

$$- \partial^2 A_\mu = J_\mu + (1/2) \psi \gamma^0 eq\gamma_\mu \psi + \dots. \tag{2.27}$$

Again, we use a special notation, A_ψ, for the vector potential under the influence of ψ to distinguish it from the noninteracting field A_μ. Then

$$A_\psi = A + \int D_+ (1/2) \psi \gamma^0 eq\gamma \psi + \dots, \tag{2.28}$$

and

$$w = \int (dx) \left[(1/2) J_\mu A^\mu + (1/2) \eta \gamma^0 \psi \right] + (1/2) \int (dx) \psi \gamma^0 eq\gamma A \psi$$
$$+ (1/2) \int (dx)(dx')(1/2)(\psi \gamma^0 eq\gamma^\mu \psi)(x) D_+(x - x')(1/2)$$
$$\times (\psi \gamma^0 eq\gamma_\mu \psi)(x')$$
$$+ (1/2) \int (dx)(dx') \psi(x) \gamma^0 eq\gamma A(x) G_+(x - x') eq\gamma A(x') \psi(x')$$
$$+ \dots, \tag{2.29}$$

in which the interactions are now made explicit. The first term describes the noninteracting system, the second term is the primitive interaction, the third term describes $e^- - e^-$ or $e^- - e^+$ scattering, and the fourth term describes electron-photon scattering or pair annihilation. The infinite series in (2.29) is a sequence of increasingly elaborate interaction skeletons. Later terms in this series do not contain modifications of earlier ones.

It should be emphasized that the iterated solution is a classification of processes in terms of increasing degree of complexity. It is not a perturbation expansion. The physical electron mass m, and the physical electron charge e, which are identified originally under specific physical circumstances, will never change their significance when the class of phenomena under examination is enlarged.

2.3   Calculation of Matrix Elements

To show how to use the structure (2.29) to calculate a particular matrix element, let us consider Compton scattering as an example. Write

$$\psi = \psi_1 + \psi_2, \ A = A_1 + A_2, \tag{2.30}$$

where ψ_2 and A_2 are the fields of weak emission sources, and ψ_1 and A_1 are the fields of weak detection sources. The experimental arrangement is

FIGURE 2.5

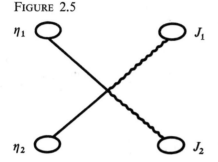

η_1 ⬤   ⬤ J_1

η_2 ⬤   ⬤ J_2

shown in Fig. 2.5. When the substitutions (2.30) are made, in (2.29), Compton scattering is described by terms involving $\psi_1\psi_2 A_1 A_2$, which are

$$w = \int (dx)(dx') \left[\psi_1(x)\gamma^0 eq\gamma A_1(x) G_+(x-x') eq\gamma A_2(x')\psi_2(x') \right.$$
$$\left. + \psi_1(x)\gamma^0 eq\gamma A_2(x) G_+(x-x') eq\gamma A_1(x')\psi_2(x') \right]. \tag{2.31}$$

In these two terms, the roles of initial and final photon are reversed. Therefore w is invariant under the interchange $A_1 \leftrightarrow A_2$. This symmetry, so called crossing symmetry, is automatically incorporated in the source theory, and is simply a consequence of particle statistics.

Now, the interaction region is far away from the sources, and since the interaction certainly occurs later in time than the emissions and earlier than the detections, we have, in the interaction region,

$$\psi_2(x) = \sum_{p\sigma q} \sqrt{d\omega_p 2m} \, u e^{ipx} i\eta_{2p\sigma q},$$

$$A_2{}^\mu(x) = \sum_{k\lambda} \sqrt{d\omega_k} \, e^\mu{}_{k\lambda} e^{ikx} iJ_{2k\lambda},$$

$$\psi_1(x)\gamma^0 = \sum_{p\sigma q} i\eta_{1p\sigma q}{}^* \sqrt{d\omega_p 2m} \, u^* \gamma^0 e^{-ipx},$$

$$A_1{}^\mu(x) = \sum_{k\lambda} iJ_{1k\lambda}{}^* \sqrt{d\omega_k} \, e^\mu{}_{k\lambda}{}^* e^{-ikx}. \tag{2.32}$$

The relevant part of the vacuum amplitude is

$$\langle 0_+ | 0_- \rangle^w \to iw, \tag{2.33}$$

where w is the expression given by (2.31). On the other hand,

$$\langle 0_+ | 0_- \rangle^{\eta J} = \sum_{\{n\},\{n'\}} \langle 0_+ | \{n\} \rangle^{\eta_1 J_1} \langle \{n\} | \{n'\} \rangle \langle \{n'\} | 0_- \rangle^{\eta_2 J_2},$$

in which there is a contribution of the form

$$\sum \langle 0_+ | 1_{(p\sigma q)_1} 1_{(k\lambda)_1} \rangle^{\eta_1 J_1} \langle 1_{(p\sigma q)_1} 1_{(k\lambda)_1} | 1_{(p\sigma q)_2} 1_{(k\lambda)_2} \rangle$$

$$\times \langle 1_{(p\sigma q)_2} 1_{(k\lambda)_2} | 0_+ \rangle^{\eta_2 J_2}. \tag{2.34}$$

But we know that

$$\langle 0_+ | 1_{(p\sigma q)_1} 1_{(k\lambda)_1} \rangle^{\eta_1 J_1} = i J_{k_1 \lambda_1}{}^* i \eta_{1 p_1 \sigma_1 q_1}{}^*,$$

$$\langle 1_{(p\sigma q)_2} 1_{(k\lambda)_2} | 0_- \rangle^{\eta_2 J_2} = i J_{k_2 \lambda_2} i \eta_{2 p_2 \sigma_2 q_2}. \tag{2.35}$$

Substituting the expressions (2.32) into (2.33), we identify, with the aid of (2.34) and (2.35),

$$\langle 1_{(p\sigma q)_1} 1_{(k\lambda)_1} | 1_{(p\sigma q)_2} 1_{(k\lambda)_2} \rangle$$

$$= ie^2 \sqrt{d\omega_{p_1} d\omega_{p_2} (2m)^2 d\omega_{k_1} d\omega_{k_2}} (2\pi)^4 \delta(p_1 + k_1 - p_2 - k_2)$$

$$\times \left[u_1{}^* \gamma^0 \gamma e_1{}^* \frac{1}{\gamma P + m} \gamma e_2 u_2 + u_1{}^* \gamma^0 \gamma e_2 \frac{1}{\gamma(P - k_1 - k_2) + m} \gamma e_1{}^* u_2 \right].$$

$$(P = p_1 + k_1 = p_2 + k_2) \tag{2.36}$$

Standard procedures can now be followed to calculate the differential cross section, etc.

2.4   Two-Particle Exchange

The first stage of dynamical description introduces an interaction skeleton, which brings various processes into existence, but is not yet a realistic theory. We reach the second stage which is a realistic, and for many purposes accurate description by taking into account two-particle exchanges.

The introduction of the primitive interaction implies that extended η sources can interact by exchanging an electron, or by the exchange of an electron and a photon (Fig. 2.6). The propagation function $G_+(x-x')$

FIGURE 2.6

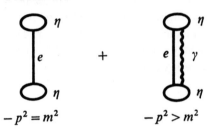

$$-p^2 = m^2 \qquad\qquad -p^2 > m^2$$

will be modified to account for the fact that mass values other than $-p^2 = m^2$ can now be exchanged. The new propagation function $\bar{G}_+(x-x')$ must have the form

$$\bar{G}_+ = G_+ + \dots,$$

where the additional terms refer to higher mass exchanges. If the sources cannot supply sufficient mass, then $\bar{G}_+$ effectively reduces to G_+, as it should, since this is the situation under which G_+ was originally introduced. The primitive interaction also contains the possibility that a photon source with $-k^2 > (2m)^2$ creates an electron-positron pair. Two such sources interact by exchanging a photon, or an electron-positron pair (Fig. 2.7), and the

FIGURE 2.7

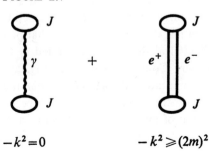

$$-k^2 = 0 \qquad\qquad -k^2 \geqslant (2m)^2$$

propagation function $D_+(x-x')$ will be modified. The new propagation function $\bar{D}_+(x-x')$, as in the case of electron sources, must have the form

$$\bar{D}_+ = D_+ + \dots,$$

where the additional terms refer to higher mass exchanges. These substitutions $(G_+ \to \bar{G}_+$ and $D_+ \to \bar{D}_+)$ must occur in all processes that can be analyzed into single particle exchanges. However, this does not exhaust the implications of two-particle exchanges. Idealized sources only partially describe realistic processes and additional effects that are characteristic of the specific interaction will appear. The simplest illustration is supplied by the primitive interaction itself. The latter describes an extended J as the source of an electron-positron pair, emitted under circumstances in which the particles have no opportunity to interact. In the next stage we enlarge the physical circumstances to permit such interactions to occur. The electron-positron interaction is described by the third term of (2.29) which can be symbolized as $(\psi\psi)D_+(\psi\psi)$. Now if we write

$$\psi = \psi_1 + \psi_2,$$

there are two terms which describe the e^+e^- scattering, $(\psi_1\psi_2)D_+(\psi_1\psi_2)$ and $(\psi_1\psi_1)D_+(\psi_2\psi_2)$. These are represented in Fig. 2.8. The first term is

FIGURE 2.8

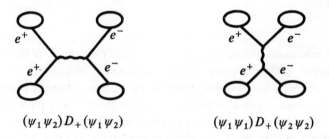

$$(\psi_1\psi_2)D_+(\psi_1\psi_2) \qquad\qquad (\psi_1\psi_1)D_+(\psi_2\psi_2)$$

ordinary Coulomb scattering and the second term describes pair annihilation and subsequent pair recreation. If now the electron and positron created by an extended photon source are permitted to interact, one recognizes that the second mechanism just described is the one that produces the modification $D_+ \to \bar{D}_+$. The dynamical details are suggested in Fig. 2.9(a). The first mechanism, the ordinary Coulomb scattering [Fig. 2.9(b)] leads to a new phenomenon. The quantitative theory shows that the effect is represented by the alteration of the electromagnetic properties of the electron, namely, the introduction of an electric form factor, and an additional magnetic moment, with its form factor.

In order to get a complete picture of the second stage of description, one must also consider multiple two-particle exchanges. The pair creation from an extended photon source is viewed as the conversion of a virtual photon

FIGURE 2.9

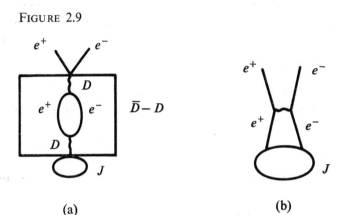

(a)   (b)

FIGURE 2.10

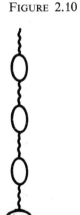

into an electron-positron pair. Now the process of recombination leads again to a virtual photon. And the whole process can be repeated indefinitely (Fig. 2.10). When we consider electron-positron scattering, the final particles may again interact, and the whole process can be indefinitely repeated. The new physical possibility the latter implies is bound electron-positron states, the positronium particles. Thus physics becomes more realistic, although it need not be completely accurate, at this level of description.

2.5   Modified Photon Propagation Function

I would like now to give a general discussion of the modified photon propagation function $\bar{D}_+$. The coupling between two photon sources is produced by the exchange of a photon, or any multiparticle exchange, with $-k^2 = M^2 > 0$. Let's write the complete contribution of the latter as

$$i\int (dx)(dx')J_1{}^{\mu}(x)a(x-x')J_{2\mu}(x'), \tag{2.37}$$

which is to be added to $\int (dx)(dx')J_1{}^{\mu}(x)D_+(x-x')J_2{}^{\mu}(x')$.
A more complicated tensor structure cannot appear, in virtue of the general property of a photon source,

$$\partial_\mu J^\mu(x) = 0.$$

Relativistic invariance requires that $a(x-x')$ be a scalar function, and we can write

$$a(x-x') = \int\limits_{k^0>0} \frac{(dk)}{(2\pi)^3} e^{ik(x-x')}a(-k^2), \tag{2.38}$$

where k^0 is the energy liberated by the source. Now, $(-k^2 = M^2)$

$$\frac{(dk)}{(2\pi)^3} = \frac{(d\mathbf{k})}{(2\pi)^3}\cdot\frac{d(k^0)^2}{2k^0} = d\omega_k dM^2, \tag{2.39}$$

and (2.37) becomes

$$\int dM^2 a(M^2)\int (dx)(dx')J_1{}^{\mu}(x)i\int d\omega_k e^{ik(x-x')}J_{2\mu}(x'). \tag{2.40}$$

One recognizes

$$i\int d\omega_k e^{ik(x-x')} = \Delta_+(x-x', M^2)$$

for the special circumstances $x^0 > x^{0\prime}$ under consideration. Thus, inclusion of multiparticle exchange gives the modified photon propagation function

$$\bar{D}_+(x-x') = D_+(x-x') + \int dM^2 a(M^2)\Delta_+(x-x', M^2). \tag{2.41}$$

Alternatively, it has the general momentum-space representation

$$\bar{D}_+(x-x') = \int \frac{(dk)}{(2\pi)^4} e^{ik(x-x')}\bar{D}_+(k), \tag{2.42}$$

with

$$\bar{D}_+(k) = \frac{1}{k^2 - i\varepsilon} + \int dM^2 \frac{a(M^2)}{k^2 + M^2 - i\varepsilon}. \tag{2.43}$$

There are general physical restrictions on the function $a(M^2)$ from vacuum persistence probability considerations. For a weak source, we have

$$\langle 0_+|0_-\rangle^J = 1 + (i/2)\int (dx)(dx')J^\mu(x)\bar{D}_+(x-x')J_\mu(x')$$

$$= 1 + (i/2)\int \frac{(dk)}{(2\pi)^4} J^\mu(k)^*\bar{D}_+(k)J_\mu(k), \tag{2.44}$$

and

$$|\langle 0_+|0_-\rangle^J|^2 = 1 - \int \frac{(dk)}{(2\pi)^4} J^\mu(k)^*J_\mu(k) \, \text{Im} \, \bar{D}_+(k). \tag{2.45}$$

If the source has only space-like momenta, $k^2 > 0$, no particle creation is possible and the vacuum state must persist, that is, $|\langle 0_+|0_-\rangle^J|^2 = 1$. Therefore

$$k^2 > 0: \text{Im} \, \bar{D}_+(k) = \int dM^2 \frac{\text{Im} \, a(M^2)}{k^2 + M^2} = 0. \tag{2.46}$$

A simple theorem on Fourier transforms (for this purpose, it is convenient to introduce the new variables $k^2 = e^x$, $M^2 = e^y$, where x and y range from $-\infty$ to $+\infty$) shows that

$$\text{Im} \, a(M^2) = 0. \tag{2.47}$$

Next, consider the situation in which the source contains only time-like momenta. Since

$$(1/\pi) \, \text{Im} \, \bar{D}_+(k) = \delta(k^2) + \int dM^2 a(M^2)\delta(k^2 + M^2), \tag{2.48}$$

and

$$\int \frac{(dk)}{(2\pi)^4} \pi\delta(k^2 + M^2) = \int \frac{(d\mathbf{k})}{(2\pi)^3} \cdot \frac{1}{2k^0} = \int d\omega_k, \tag{2.49}$$

we get

$$|\langle 0_+|0_-\rangle^J|^2 = 1 - \int d\omega_k J^\mu(k)^*J_\mu(k)/_{k^2=0}$$

$$- \int dM^2 a(M^2)\int d\omega_k J^\mu(k)^*J_\mu(k)/_{-k^2=M^2}. \tag{2.50}$$

In the rest frame of k^μ, $M \neq 0$, we have $J^0(k) = 0$ and $J^\mu(k)^*J_\mu(k) = |\mathbf{J}(k)|^2 > 0$. Therefore it is necessary that

$$a(M^2) \geqslant 0, \tag{2.51}$$

that is, $a(M^2)$ must be real and non-negative.

We now consider a simple mechanism of single pair exchange (Fig. 2.7). The pair creation by an extended photon source is described by the primitive interaction. From its contribution to the vacuum amplitude

$$i(1/2)\int (dx)\psi(x)\gamma^0\gamma^\mu eqA_\mu(x)\psi(x) \tag{2.52}$$

it is seen that A_μ appears as an effective electron-positron source. Comparing with the description of the noninteracting propagation of two particles

$$\exp\left[i\int (dx)\psi(x)\gamma^0\eta(x)\right] \to -(1/2)\int (dx)(dx')\psi(x)\gamma^0\eta(x)$$
$$\times \eta(x')\gamma^0\psi(x'), \qquad (2.53)$$

one identifies

$$\eta(x)\eta(x')/_{\text{eff}} = -ieq\gamma^\mu\gamma^0 A_\mu(x)\delta(x-x'), \qquad (2.54)$$

where the left-hand side is regarded as a matrix. The consideration of two such effective electron-positron sources give the following contribution to the vacuum amplitude:

$$-(1/2)\int (dx)(dx')A_1{}^\mu(x)A_2{}^\nu(x')\,\text{Tr}\left[eq\gamma_\mu G_+(x-x')eq\gamma_\nu\right.$$
$$\left.\times G_+(x'-x)\right], \qquad (2.55)$$

when the substitution (2.54) is made for $\eta_1(x)\eta_1(x')$ and $\eta_2(x)\eta_2(x')$ in the expression

$$\exp\left[i\int (dx)(dx')\eta_1(x)\gamma^0 G_+(x-x')\eta_2(x')\right]$$
$$\to -(1/2)\left[\int (dx)(dx')\eta_1(x)\gamma^0 G_+(x-x')\eta_2(x')\right]^2.$$

Under the well-defined causal circumstances $(x^0 > x^{0\prime})$, the two electron propagation functions have the form

$$G_+(x-x') = i\int d\omega_p e^{ip(x-x')}(m-\gamma p), \qquad (2.56)$$

$$G_+(x'-x) = i\int d\omega_{p'} e^{ip'(x-x')}(m+\gamma p'). \qquad (2.57)$$

Now (2.55) becomes

$$-(1/2)\int d\omega_p d\omega_{p'} \frac{J_1{}^\mu(-k)J_2{}^\nu(k)}{(k^2)^2}$$
$$\times \text{Tr}\left[eq\gamma_\mu(m-\gamma p)eq\gamma_\nu(-m-\gamma p')\right], \qquad (2.58)$$

where

$$k = p + p', \quad .$$

is the total momentum liberated by the source. We transfer our attention to the total momentum exchange, k, by introducing the unit factor $(-k^2 = M^2)$

$$(2\pi)^3 \int dM^2 d\omega_k \delta(p+p'-k). \qquad (2.59)$$

Thus, (2.58) reduces to

$$-(e^2/2)\int dM^2 d\omega_k J_1{}^\mu(-k)J_2{}^\nu(k)(1/M^2)^2 I_{\mu\nu}(k), \qquad (2.60)$$

where

$$I_{\mu\nu}(k) = (2\pi)^3 \int d\omega_p d\omega_{p'} \delta(p+p'-k)$$
$$\times \text{Tr}\left[q\gamma_\mu(m-\gamma p)q\gamma_\nu(-m-\gamma p')\right]. \qquad (2.61)$$

A simple evaluation in the rest frame of k gives the result

$$I_{\mu\nu}(k) = \left(g_{\mu\nu} + \frac{k_\mu k_\nu}{M^2}\right)\frac{1}{6\pi^2}M^2\sqrt{1-\left(\frac{2m}{M}\right)^2}\left(1+\frac{2m^2}{M^2}\right), \qquad (2.62)$$

in which the $k_\mu k_\nu$ term will not contribute, since the current J_μ is conserved. With the definition

$$a(M^2) = \frac{\alpha}{3\pi}\cdot\frac{1}{M^2}\sqrt{1-\left(\frac{2m}{M}\right)^2}\left(1+\frac{2m^2}{M^2}\right), \quad M > 2m, \qquad (2.63)$$

where $\alpha = \dfrac{e^2}{4\pi}$ is the physical fine-structure constant, we get the contribution to the vacuum amplitude

$$-\int dM^2 a(M^2)d\omega_k J_1{}^\mu(-k)J_{2\mu}(k)$$
$$= i\int dM^2 a(M^2)\int (dx)(dx')J_1{}^\mu(x)i\int d\omega_k e^{ik(x-x')}J_{2\mu}(x'). \qquad (2.64)$$

Again we recognize that $\Delta_+(x-x', M^2)$ is the appropriate generalization for the $id\omega_k$ integral. The modified photon propagation function appears as

$$\bar{D}_+(k) = \frac{1}{k^2-i\varepsilon} + \int dM^2 \frac{a(M^2)}{k^2+M^2-i\varepsilon},$$

which is a particular realization of the general form (2.43). The function $a(M^2)$ given by (2.63) is real and non-negative. For $M \gg 2m$, the integral behaves like

$$\int \frac{dM^2}{M^2}\frac{1}{k^2+M^2}$$

and there is no question about its existence.

2.6   Electromagnetic Form Factors

I would like now to discuss the modifications of the electromagnetic
properties of the electron, following the same pattern of discussion as in the
simpler case of the modified photon propagation function. Firstly we
consider the general structure that one should expect on the basis of kinematics and causality, and secondly we will calculate a specific mechanism
within that frame work.

FIGURE 2.11

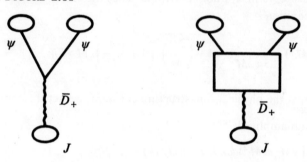

We are interested in the interaction modification of the pair creation
by an extended photon source. These processes can be classified into two
types. Those in which all modifications of D_+ into $\bar{D}_+$ take place and then
the emission of the final pair occurs in one act, and all others which cannot
be so analyzed. The separation of these two types of processes are represented
pictorially in Fig. 2.11. Correspondingly, we write the vacuum amplitude
as two terms, for weak sources,

$$i\int (dx)(1/2)\psi(x)\gamma^0\gamma^\mu eq\psi(x)\bar{A}_\mu(x)$$

$$-\int (dx)(dx')(d\xi)(1/2)\psi(x)\gamma^0 eqK^\mu(xx',\xi)\psi(x')\bar{A}_\mu(\xi), \qquad (2.65)$$

where $\bar{A}$ is the vector potential constructed from the photon source J with
the aid of the modified propagation function $\bar{D}_+$. The momentum space
form of the second term is

$$-\frac{1}{2}\int \frac{(dk)}{(2\pi)^4} d\omega_p d\omega_{p'} \eta(-p)\gamma^0(m-\gamma p)eqK^\mu(p,p',k)$$

$$\times (-m-\gamma p')\eta(-p')\bar{A}_\mu(k). \qquad (2.66)$$

Translational invariance requires the following structure for K^μ:

$$\frac{1}{2\pi}K^\mu(p,p',k) = \delta(p+p'-k)(2\pi)^4 K^\mu(p-p',k). \qquad (2.67)$$

Consider the general structure of $K^\mu(p-p', k)$. Now

$$-k^2 = M^2,$$

$$k(p-p') = p^2 - p'^2 = 0, \ (-p^2 = -p'^2 = m^2)$$

$$(p-p')^2 + k^2 = -4m^2,$$

therefore, M^2 is the only scalar that can be constructed from the momentum vectors p, p' and k. The possible vectors are γ^μ, k^μ and $(p-p')^\mu$. Suppose we use Lorentz gauge, so that

$$k^\mu \bar{A}_\mu(k) = 0.$$

Furthermore,

$$(p-p')^\mu = -(1/2)\{\gamma^\mu, \gamma p - \gamma p'\}.$$

In virtue of the projection matrices $(m - \gamma p)$ and $(-m - \gamma p')$ in (2.66), when γp appears as a left factor (denoted as $\overleftarrow{\gamma p}$), it can be replaced by $-m$, and similarly a right factor of $\gamma p'$ (denoted as $\overrightarrow{\gamma p'}$) can be replaced by $+m$. Then

$$\overrightarrow{\gamma p} = \gamma k - \overrightarrow{\gamma p'} \to \gamma k - m,$$

$$\overleftarrow{\gamma p'} = \gamma k - \overleftarrow{\gamma p} \to \gamma k + m.$$

Therefore, the only possible structure for $K^\mu(p-p', k)$ is

$$K^\mu(p-p', k) = \gamma^\mu K_1(M) + (1/2)[\gamma^\mu, \gamma k]K_2(M),$$

or

$$K^\mu(p-p', k) = \gamma^\mu K_1(M) - i\sigma^{\mu\nu}k_\nu K_2(M). \tag{2.68}$$

Since there is no individual reference to p and p', (2.66) can be put back in coordinate space in the form

$$i(1/2)\int (dx)\psi(x)\gamma^0 eq\gamma^\mu \psi(x)e^{ikx}dM^2 K_1(M)id\omega_k \bar{A}_\mu(k)$$

$$+ i(1/2)\int (dx)\psi(x)\gamma^0 eq(1/2)\sigma^{\mu\nu}\psi(x)e^{ikx}dM^2$$

$$\times K_2(M)id\omega_k \bar{F}_{\mu\nu}(k), \tag{2.69}$$

where

$$\bar{A}_\mu(k) = \int (dx')e^{-ikx'}\bar{A}_\mu(x'), \tag{2.70}$$

and

$$\bar{F}_{\mu\nu}(k) = i[k_\mu \bar{A}_\nu(k) - k_\nu \bar{A}_\mu(k)] \tag{2.71}$$

is the field strength tensor.

We are discussing an addition to $\int (dx)(1/2)\psi\gamma^0 eq\gamma^\mu\psi\bar{A}_\mu$ which contains the primitive interaction and a description of how the gauge variance of A (or $\bar{A}$) is tied to ψ. Therefore, one requires all additional effects to be gauge invariant. The second term in (2.69) is gauge invariant by itself, and the first term can also be so written $\left[k_\mu\bar{A}^\mu(k)=0\right]$,

$$\bar{A}_\mu(k) = -\frac{1}{M^2}k^2\bar{A}_\mu(k)$$

$$= \frac{1}{M^2}k^\nu\left[k_\mu\bar{A}_\nu(k)-k_\nu\bar{A}_\mu(k)\right]. \tag{2.72}$$

The causal situation under consideration is clear cut. The source produces a virtual photon which converts into an electron-positron pair close to the source. Then various interactions take place at a later time. Thus $x^0 > x^{0'}$, and again we can introduce $\Delta_+(x-x', M^2)$ as the appropriate space-time generalization of $i\int d\omega_k e^{ik(x-x')}$.

The complete structure of (2.65) is then

$$i\int (dx)(dx')(1/2)\psi(x)\gamma^0 eq\gamma^\mu\psi(x)F_1(x-x')\bar{A}_\mu(x')$$

$$+ i\int (dx)(dx')(1/2)\psi(x)\gamma^0\frac{eq}{2m}(1/2)\sigma^{\mu\nu}\psi(x)\mu'$$

$$\times F_2(x-x')\bar{F}_{\mu\nu}(x'), \tag{2.73}$$

where, using Fourier transforms,

$$F_1(k) = 1 - k^2\int dM\,\frac{\phi_1(M)}{k^2+M^2-i\varepsilon},\ F_1(0) = 1, \tag{2.74}$$

$$F_2(k) = \int dM\,\frac{\phi_2(M)}{k^2+M^2-i\varepsilon},\ F_2(0) = 1,$$

and

$$\phi_1(M) = \frac{2K_1(M)}{M},\ \frac{\mu'}{2m}\,\phi_2(M) = 2MK_2(M).$$

Here $F_1(k)$ and $F_2(k)$ are, respectively, the electric and magnetic form factor. And μ' in (2.73) is the anomalous magnetic moment.

Let us now carry out a calculation of a specific mechanism. Consider the effect of the Coulomb interaction of the electron-positron pair emitted by an extended photon source. As we have discussed [see (2.54) in particular], the primitive interaction describes an extended photon source as an effective electron-positron source. Inserting this effective two-particle source into the electron-positron scattering description, one gets a contribution to the vacuum amplitude given by

$$(1/2)\int (dx)(dx')(d\xi)\psi(x)\gamma^0 eq\gamma^\mu G_+(x-\xi)$$

$$\times eq\gamma A(\xi)G_+(\xi-x')eq\gamma_\mu\psi(x')D_+(x-x').\qquad (2.75)$$

FIGURE 2.12

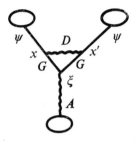

The situation is described in Fig. 2.12. The causal arrangement is such that $x^0 > \xi^0$, and $x^{0'} > \xi^0$. The evaluation is performed without difficulty in the center-of-mass frame of the two-particle system. Only one special point should be mentioned. It is the integration over all scattering angles. The long-range Coulomb interaction gives infinity in the forward direction. This difficulty can be removed by using a screened Coulomb potential, i.e. $(1/r)e^{-\mu r}$, which is equivalent to the unphysical use of a small photon mass μ. It is clear that this is simply a version of the infra-red problem. This problem is easily resolved by asking the proper physical question, that is, soft photon emission processes must also be considered. The result is expressed in the general framework of (2.74) as

$$\mu' = \frac{\alpha}{2\pi},$$

$$\phi_2(M) = \frac{(2m)^2}{M}\frac{1}{\sqrt{1-\left(\frac{2m}{M}\right)^2}},\quad M > 2m,$$

$$\phi_1(M) = \frac{\alpha}{\pi}\frac{1}{M}\frac{1}{\sqrt{1-\left(\frac{2m}{M}\right)^2}}\left[\left(1-\frac{2m^2}{M^2}\right)\log\frac{M^2-4m^2}{\mu^2}\right.$$

$$\left. -\frac{3}{2}+\left(\frac{2m}{M}\right)^2\right],\quad M > 2m.\qquad (2.76)$$

It should be noticed that all the integrals exist.

The result that we have just obtained describes the electromagnetic properties of a free electron. For the Lamb-shift calculation, on the other hand, one has to deal with a bound electron with $p^2 + m^2 \neq 0$. In this case one can use J to represent the charge distribution of the nucleus, which produces its effect through A. The basic problem is then the multiple scattering of an

FIGURE 2.13

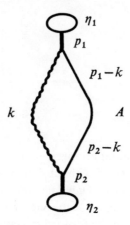

electron in a Coulomb field. To obtain the radiative interaction modification one considers the scattering problem using extended electron sources to produce a well-defined causal situation. The extended electron source η_2 emits a virtual electron of momentum p_2 which cannot last very long and converts into a real photon of momentum k and a real electron of momentum $p_2 - k$. The electron moves to the nuclear potential and is scattered. The final electron of momentum $p_1 - k$ and the photon recombine into a virtual electron of momentum p_1 which again cannot last very long and is detected by the extended electron source η_1. The causal sequence is schematically represented in Fig. 2.13. The calculational advantage of this process is that the integration over all possible photon momenta k is limited by kinematics,

$$k^2 = 0, \ (p_2 - k)^2 + m^2 = 0, \ (p_1 - k)^2 + m^2 = 0, \tag{2.77}$$

and only one parameter is left, which appears as an angle in a certain coordinate system (say, $\mathbf{p}_2 = 0$, $p_2{}^0 = M_2$). The result of this calculation is a contribution to the vacuum amplitude of the form

$$i \int \frac{(dp_1)}{(2\pi)^3} \frac{(dp_2)}{(2\pi)^3} \, \psi_1(-p_1)\gamma^0(-eq)A_\mu(p_1 - p_2)$$

$$\times [g_1\gamma^\mu + g_2 i\sigma^{\mu\nu}(p_1 - p_2)_\nu]\psi_2(p_2), \tag{2.78}$$

where g_1 and g_2 are invariant functions of the three scalars, $-p_1{}^2 = M_1{}^2$, $-p_2{}^2 = M_2{}^2$, and $(p_1 - p_2)^2$.

We now write

$$\psi_1(-p_1) = \int (dx_1)\psi_1(x_1)e^{ip_1x_1},$$
$$\psi_2(p_2) = \int (dx_2)e^{-ip_2x_2}\psi_2(x_2),$$
$$A_\mu(p_1 - p_2) = \int (d\xi)e^{-i(p_1-p_2)\xi}A_\mu(\xi). \tag{2.79}$$

The well-defined causal arrangement guarantees that $x_1{}^0 > \xi^0$, and $\xi^0 > x_2{}^0$. On writing

$$\frac{(dp_1)}{(2\pi)^3} = d\omega_{p_1}dM_1{}^2, \quad \frac{(dp_2)}{(2\pi)^3} = d\omega_{p_2}dM_2{}^2, \tag{2.80}$$

we find that the typical structure in space-time representation is

$$\int (dx_1)(dx_2)(d\xi)dM_1{}^2 dM_2{}^2 \psi_1(x_1)\Delta_+(x_1 - \xi, M_1{}^2)$$
$$\times A(\xi)\Delta_+(\xi - x_2, M_2{}^2)\psi_2(x_2), \tag{2.81}$$

in which $\Delta_+(\xi - x_2, M_2{}^2)$ and $\Delta_+(x_1 - \xi, M_1{}^2)$ describe, respectively, the transmission of mass M_2 from the extended source η_2 to the nuclear potential $A(\xi)$, and transmission of mass M_1 from $A(\xi)$ to the extended source η_1. After the space-time generalization, we return to the momentum-space representation,

$$-i \int \frac{(dp_1)}{(2\pi)^4} \frac{(dp_2)}{(2\pi)^4} \psi_1(-p_1)\gamma^0 [g_1\gamma^\mu(-eq)A_\mu(p_1 - p_2)$$
$$+ g_2 eq(1/2)\sigma^{\mu\nu}F_{\mu\nu}(p_1 - p_2)]\psi_2(p_2)$$
$$\times \int dM_1{}^2 dM_2{}^2 \left[\frac{1}{p_1{}^2 + M_1{}^2 - i\varepsilon} \right.$$
$$\left. \cdot \frac{1}{p_2{}^2 + M_2{}^2 - i\varepsilon} - f((p_1 - p_2)^2, M_1{}^2, M_2{}^2) \right], \tag{2.82}$$

in which $p_1{}^2$ and $p_2{}^2$ are not restricted to $-p_1{}^2 = -p_2{}^2 = m^2$. The additional term $f((p_1 - p_2)^2, M_1{}^2, M_2{}^2)$ appears here because the discussion by means of the well-defined causal situation effectively studies the dependence upon $p_1{}^2$ and $p_2{}^2$, but leaves undetermined any contribution that involves only the third invariant $(p_1 - p_2)^2$. The f term can be removed by subtracting the already known result for $-p_1{}^2 = -p_2{}^2 = m^2$. Despite the way (2.82) has been written, this is to be performed separately for the γA and the σF terms, and turns out to be unnecessary for the latter.

3. STRONG AND WEAK INTERACTIONS

3.1 Low-Energy $\pi+N$ System

In electrodynamics the very intuitive idea that an accelerated charge radiates enables us to elaborate the whole dynamics of electrons and photons, beginning with the primitive interaction. Nothing so simple is available in strong and weak dynamics. We shall exhibit another facet of the source formalism by using it to search for empirical regularities, with a minimum use of speculative hypotheses.

The essential contribution of the source theory is the possibility of presenting the phenomenological description of a physical system in the general form

$$\langle 0_+ | 0_- \rangle^S = \exp\{i\int (dx)[S\chi + \mathscr{L}(\chi)]\}, \tag{3.1}$$

where χ are the fields corresponding to the sources S, and $\mathscr{L}(\chi)$ is the Lagrange function of the system. The latter is not a local function of the fields, but may be so represented under limited physical circumstances. Consider for illustration the $\pi+N$ system at low energy. The pion is an isotopic triplet $(T=1)$, and the nucleon is an isotopic doublet $(T=(1/2))$. One also has to distinguish a nucleon from an antinucleon. This distinction is conveyed by the use of a complex source. We will use the particle symbols to represent the corresponding fields. For the $\pi+N$ system we have then

$$S\chi \to K \cdot \pi + \bar\eta N + \bar N \eta, \tag{3.2}$$

where

$$\bar{N} = N^* \gamma^0, \tag{3.3}$$

and

$$\mathscr{L} = \mathscr{L}_\pi + \mathscr{L}_N + \mathscr{L}_{\pi N}, \tag{3.4}$$

$$\mathscr{L}_\pi = -(1/2)\left[(\partial_\mu \pi)^2 + m_\pi^2 \pi^2\right], \tag{3.5}$$

$$\mathscr{L}_N = -\bar{N}(\gamma^\mu (1/i)\partial_\mu + M)N. \tag{3.6}$$

Some significant interaction terms for the low-energy $\pi + N$ system are

$$\mathscr{L}_{\pi N} = \frac{f}{m_\pi} \bar{N}\gamma^\mu i\gamma_5 \tau N \cdot \partial_\mu \pi$$

$$+ \left(\frac{f_0}{m_\pi}\right)^2 \bar{N}\gamma^\mu \tau N \cdot \partial_\mu \pi \times \pi. \tag{3.7}$$

If we consider the exchange of a virtual π, the first term gives the long-range behavior of nuclear forces. If the exchange of a virtual nucleon is considered, it describes the low-energy limit of p-wave π–N scattering. The value of f obtained in this way is

$$f = 1.01 \pm 0.01, \tag{3.8}$$

or $\dfrac{f^2}{4\pi} \sim 0.08$. The second term is introduced phenomenologically to account for the low-energy s-wave π-N scattering. The experimental data are

$$m_\pi(a_{(1/2)} - a_{(3/2)}) = 0.292 \pm 0.02,$$

$$m_\pi(2a_{(3/2)} + a_{(1/2)}) = -0.035 \pm 0.012, \tag{3.9}$$

where $a_{(1/2)}$ and $a_{(3/2)}$ are, respectively, the scattering amplitudes for the $\pi + N$ system in the $T = (1/2)$ and $T = (3/2)$ states. The f_0^2 term alone gives $2a_{(3/2)} + a_{(1/2)} = 0$. But the f term also gives a small s-wave contribution

$$m_\pi(2a_{(3/2)} + a_{(1/2)}) = -3\frac{f^2}{4\pi} \frac{\dfrac{m_\pi}{M}}{1 + \dfrac{m_\pi}{M}} \simeq -0.03, \tag{3.10}$$

where $\dfrac{m_\pi}{M} \simeq \dfrac{1}{6.9}$. The result is consistent with experiment. The relation

$$m_\pi(a_{(1/2)} - a_{(3/2)}) = \frac{3}{2\pi} f_0^2 \frac{1}{1 + \dfrac{m_\pi}{M}}, \tag{3.11}$$

gives

$$f_0 = 0.84 \pm 0.03. \tag{3.12}$$

The two parameters f and f_0 are the information extracted from the low-energy strong interaction phenomenology of the $\pi + N$ system.

3.2   Partial Symmetry. Chiral Invariance

The Lagrange function (3.4) is invariant under the isotopic rotations,

$$\delta\pi = -\delta\omega \times \pi,$$

$$\delta N = (i/2)\tau \cdot \delta\omega N, \tag{3.13}$$

which form a group O_3 or SU_2. This invariance is actually only approximate, since

$$m_{\pi^0} \neq m_{\pi^{\pm}},$$

$$M_n \neq M_p.$$

Isotopic spin invariance is only a partial symmetry, abstracted by disregarding certain terms in the Lagrange function, specifically the mass differences in the present case. This is a very general feature, empirically, of strong interaction dynamics. Another significant example of partial symmetry is obtained upon the recognition that the pion mass is the smallest of all meson masses. If it were zero, then the Lagrange function

$$\mathscr{L}_\pi = -(1/2)(\partial_\mu\pi)^2$$

would have a new symmetry. It is invariant under the displacement of the π field by a constant,

$$\pi \to \pi + \varphi.$$

Does this symmetry have any significance for the πN interaction? Observe that it is also true for the f interaction term of (3.7), since this term involves only derivatives of the π field. However, it is not valid for the f_0 term. Considering an infinitesimal displacement, we obtain

$$\delta\mathscr{L}_{\pi N} = \left(\frac{f_0}{m_\pi}\right)^2 \overline{N}\gamma^\mu\tau N \cdot \partial_\mu(\pi \times \delta\varphi). \tag{3.14}$$

This structure can be compensated by introducing an isotopic-rotation like transformation on the nucleon fields

$$\delta N = i\left(\frac{f_0}{m_\pi}\right)^2 \tau \cdot \pi \times \delta\varphi N,$$

$$\delta\overline{N} = -i\left(\frac{f_0}{m_\pi}\right)^2 \overline{N}\tau \cdot \pi \times \delta\varphi. \tag{3.15}$$

The transformation (3.15) also introduces changes in the f interaction term. But they are at least cubic in the π field and its variation, and therefore lie outside the limited framework being considered. Our attention is concentrated on only those processes involving one or two π particles.

Given a set of transformations, it is essential to examine its group structure. A simple way of testing the closure property is to consider a sequence of infinitesimal transformations, say (12), and to compare with the transformations performed in opposite order, that is, $(21)^{-1}$. The group property demands that the resultant transformation, symbolized as $(12)(21)^{-1}$, be a third transformation of the general kind. Such a procedure has commonly been used to derive the angular-momentum commutation relations. Now consider the sequence of transformations

$$N' = \left[1 + i\left(\frac{f_0}{m_\pi}\right)^2 \tau \cdot \pi \times \delta_1\varphi\right]N,$$

$$N'' = \left[1 + i\left(\frac{f_0}{m_\pi}\right)^2 \tau \cdot \pi' \times \delta_2\varphi\right]N', \quad (\pi' = \pi + \delta\varphi)$$

$$= \left[1 + i\left(\frac{f_0}{m_\pi}\right)^2 \tau \cdot (\pi + \delta_1\varphi) \times \delta_2\varphi\right]$$

$$\times \left[1 + i\left(\frac{f_0}{m_\pi}\right)^2 \tau \cdot \pi \times \delta_1\varphi\right]N. \tag{3.16}$$

On comparing with the transformations in opposite order, and keeping only lowest powers of the π field, to be consistent with our restricted starting point, the final transformation connecting the two sequences of operations is an ordinary isotopic rotation

$$\delta_{[12]}N = i\left(\frac{2f_0}{m_\pi}\right)^2 (1/2)\tau \cdot \delta_1\varphi \times \delta_2\varphi N, \tag{3.17}$$

or

$$\delta_{[12]}\omega = \lambda^2 \delta_1\varphi \times \delta_2\varphi, \tag{3.18}$$

where

$$\lambda = \frac{2f_0}{m_\pi}. \tag{3.19}$$

The corresponding group property of the isotopic rotations is

$$\delta_{[12]}\omega = \delta_1\omega \times \delta_2\omega. \tag{3.20}$$

The relabelling

$$\delta\omega_3 = \delta\omega_{12}, \text{ cyclic}, \tag{3.21}$$

$$\lambda\delta\varphi_3 = \delta\omega_{34}, \text{ etc.}, \tag{3.22}$$

enables us to recognize the group structure of O_4 (the four-dimensional Euclidean rotation group with six parameters $\delta\omega_{12}$, $\delta\omega_{34}$, etc.). It is well-known that the group O_4 can be factored

$$O_4 = O_3 \times O_3, \tag{3.23}$$

if the parameters $\delta\omega_{12} \pm \delta\omega_{34}$, etc. are introduced. Thus, we have now two independent isotopic rotation groups with parameters

$$\delta\omega_\pm = \delta\omega \pm \lambda\delta\varphi. \tag{3.24}$$

Under a spatial reflection, $\delta\omega \to \delta\omega$, but $\delta\varphi \to -\delta\varphi$, therefore $\delta\omega_+ \leftrightarrow \delta\omega_-$. The combinations $\delta\omega_+$ and $\delta\omega_-$ thus correspond to opposite handness, or chirality, which are interchanged under a spatial reflection. We shall call the O_4 transformations in general and $\delta\varphi$ transformations in particular, chiral transformations.

Thus far, we have abstracted the chiral group from the phenomenology of the low-energy $\pi + N$ system. Let's see if we can apply the result to deduce new relations. The actual response of the Lagrange function to a constant $\delta\omega$ and $\delta\varphi$ transformation is

$$\delta\mathcal{L} = -m_\pi^2 \pi \cdot \delta\varphi. \tag{3.25}$$

Following a classic Lagrangian technique for deriving currents, we now let $\delta\omega$ and $\delta\varphi$ be arbitrary space-time functions. This gives

$$\delta\mathcal{L} + m_\pi^2 \pi \cdot \delta\varphi = -j^\mu{}_T \cdot \partial_\mu \delta\omega - \lambda k^\mu \cdot \partial_\mu \delta\varphi, \tag{3.26}$$

where

$$j^\mu{}_T = \partial_\mu \pi \times \pi + \bar{N}\gamma_\mu(1/2)\tau N + \ldots, \quad \text{(isotopic current)}$$

$$\lambda k_\mu = \partial_\mu \pi - \frac{f}{m_\pi} \bar{N}\gamma_\mu i\gamma_5 \tau N + \ldots, \quad \text{(axial current)} \tag{3.27}$$

in which we have retained only the simplest contributions. The stationary property of $\mathcal{L}$ (at points far away from sources) then gives

$$\partial_\mu j^\mu{}_T = 0,$$

$$\partial_\mu k^\mu = \frac{m_\pi^2}{\lambda} \pi = \left(\frac{m_\pi}{2f_0}\right) m_\pi^2 \pi, \tag{3.28}$$

which state the conservation of isotopic current and partial conservation of axial current, respectively. Alternatively

$$\delta\mathcal{L} + m_\pi^2 \pi \cdot \delta\varphi = -(1/2)j_+{}^\mu \cdot \partial_\mu \delta\omega_+ - (1/2)j_-{}^\mu \cdot \partial_\mu \delta\omega_-, \tag{3.29}$$

where

$$j_\pm{}^\mu = j_T{}^\mu \pm k^\mu. \tag{3.30}$$

And in particular

$$j_{+\mu} = \partial_\mu \pi \times \pi + \frac{m_\pi}{2f_0} \partial_\mu \pi + \bar{N}\gamma_\mu(1/2)\tau\left(1 - i\gamma_5 \frac{f}{f_0}\right)N + \ldots \qquad (3.31)$$

The known chiral structure of lepton currents in weak interaction suggests the use of $j_+{}^\mu$ as the corresponding $\pi + N$ weakly interacting current. This single hypothesis puts together in one package a number of hypotheses accumulated over the years. These include: the Feynman-Gell-Mann hypothesis connecting pion β-decay to nucleon vector β-decay coupling; the Goldberger-Treiman relation between pion instability and the nucleon axial vector coupling; and the nucleonic axial-vector—vector ratio

$$-\frac{G_A}{G_V} = \frac{f}{f_0} = \frac{1.01}{0.84} = 1.20 \pm 0.03 \qquad (3.32)$$

(Experimental value: 1.18 ± 0.027)

which is the Adler-Weisberger relation, as interpreted by Tomozawa and Weinberg. Later we will discuss another approach, which gives a pure number for the strong- and weak-interaction ratios. It is

$$-\frac{G_A}{G_V} = \frac{(5/3)}{\sqrt{2}} = 1.18. \qquad (3.33)$$

3.3   Non-Abelian Vector Gauge Particles, ρ and A_1

As the experimental energy increases, the assumption of locality for the Lagrange function becomes invalid. But we also begin to recognize new particles, in terms of which we can hope to regain some of the simplicity of a local description. Consider the lightest 1^- particle, the ρ meson, $T = 1$. It will be natural to build a theory in analogy to the photon, with the isotopic current replacing the electric current. But the ρ meson itself must contribute to the isotopic current, while the photon is electrically neutral. Also the non-zero mass of the ρ meson is a fundamental difference. The attitude we want to take is another example of partial symmetry. We now approach this problem with a gauge invariant requirement that would be exact in the absence of the ρ mass term.

Since the ρ meson is coupled to itself, the gauge invariance here is a more complicated kind. Consider the $\rho + \pi$ system for illustration. Take the Lagrange function as

$$\mathscr{L} = -(1/2)(D_\mu \pi)^2 - (1/2)m_\pi^2 \pi^2$$

$$-(1/4)(\partial_\mu \rho_\nu - \partial_\nu \rho_\mu + g\rho_\mu \times \rho_\nu)^2 - (1/2)m_\rho^2(\rho_\mu)^2. \tag{3.34}$$

where

$$D_\mu = \partial_\mu + g\rho_\mu \times \tag{3.35}$$

is the analogue of $\partial_\mu - ieqA_\mu$ in electrodynamics. Without the ρ mass term, $\mathscr{L}$ is invariant under the isotopic-spin gauge transformation (non-Abelian gauge transformation),

$$\delta\pi = -\delta\omega(x) \times \pi,$$

$$\delta\rho_\mu = -\delta\omega(x) \times \rho_\mu + (1/g)\partial_\mu \delta\omega(x) = (1/g)D_\mu \delta\omega(x). \tag{3.36}$$

The first term in the ρ transformation expresses the fact that the ρ meson is an isotopic-spin-1 particle, while the second term is characteristic of a gauge field. The general response of $\mathscr{L}$ is

$$\delta\mathscr{L} = -\frac{m_\rho^2}{g}\rho^\mu \cdot \partial_\mu \delta\omega(x), \tag{3.37}$$

which implies, in particular,

$$\partial_\mu \rho^\mu = 0. \tag{3.38}$$

When the nucleon is included, the relevant structure in $\mathscr{L}$ is

$$-\bar{N}\gamma^\mu((1/i)\partial_\mu - g(1/2)\tau \cdot \rho_\mu)N + \text{mass term}, \tag{3.39}$$

where N responds to the gauge transformation as

$$\delta N = i(\tau/2) \cdot \delta\omega(x)N. \tag{3.40}$$

We can find the connection with the previous approach through the ρ field equation

$$m_\rho^2 \rho_\mu + \partial^\nu(\partial_\mu\rho_\nu - \partial_\nu\rho_\mu) = g(\partial_\mu\pi \times \pi + \bar{N}\gamma_\mu(1/2)\tau N + \ldots).$$

$$= g j_{\mu T}. \tag{3.41}$$

Under the previous low-energy circumstances, the ρ meson cannot be produced as a real particle, i.e., $\partial^2 \ll m_\rho^2$, and (3.41) reduces to

$$\frac{m_\rho^2}{g} \rho_\mu \simeq j_{\mu T}, \tag{3.42}$$

so (3.37) gives the previous result. It indicates that the introduction of the ρ meson is a consistent generalization.

We can also make contact between the ρ and chiral transformations. Under the displacement

$$\delta\pi = \delta\varphi \text{ (constant)}, \tag{3.43}$$

we have

$$\delta j_{\mu T} = \partial_\mu\pi \times \delta\varphi + \ldots$$

$$= \partial_\mu(\pi \times \delta\varphi) + \ldots. \tag{3.44}$$

The low-energy connection (3.42) gives

$$\delta\rho_\mu = \frac{g}{m_\rho^2} \partial_\mu(\pi \times \delta\varphi) + \ldots, \tag{3.45}$$

and we identify the gauge parameter $\delta\omega$,

$$\delta\omega = \left(\frac{g}{m_\rho}\right)^2 \pi \times \delta\varphi. \tag{3.46}$$

The corresponding nucleon transformation (3.40) now gives

$$\delta N = \frac{i}{2}\left(\frac{g}{m_\rho}\right)^2 \tau \cdot \pi \times \delta\varphi N. \tag{3.47}$$

But we also know the appropriate chiral transformation for the nucleon; it is

$$\delta N = i\left(\frac{f_0}{m_\pi}\right)^2 \tau \cdot \pi \times \delta\varphi N. \tag{3.48}$$

The comparison of (3.47) with (3.48) implies the relation

$$\frac{1}{2}\left(\frac{g}{m_\rho}\right)^2 = \left(\frac{f_0}{m_\pi}\right)^2.$$

(3.49)

From the known values of f_0, one finds

$$\frac{g^2}{4\pi} = 3.4 \pm 0.2.$$

(3.50)

Thus, the universal coupling constant g is fixed by the low-energy pheno-menology.

Observe that the s-wave πN coupling term (the f_0^2 term) is now completely accounted for by the ρ exchange mechanism. The relevant coupling terms in $\mathscr{L}$ are

$$g\bar{N}\gamma^\mu(1/2)\tau N \cdot \rho_\mu + g\rho_\mu \cdot \partial_\mu \pi \times \pi.$$

(3.51)

Each coupling term describes an effective ρ source, and the consideration of ρ exchange between the two sources gives the contribution to the action,

$$\int (dx)(dx') g\bar{N}\gamma^\mu(1/2)\tau N(x)\left(g_{\mu\nu} - \frac{1}{m_\rho^2}\partial_\mu\partial_\nu\right)\Delta_+(x-x')_\rho$$

$$\times g\partial_\mu'\pi \times \pi(x').$$

(3.52)

For small momentum exchanges, $p^2 \ll m_\rho^2$, we have

$$\Delta_+(x-x')_\rho = \int \frac{(dp)}{(2\pi)^4}\frac{e^{ip(x-x')}}{p^2+m_\rho^2-i\varepsilon}$$

$$\to \frac{1}{m_\rho^2}\delta(x-x'),$$

(3.53)

and (3.52) becomes

$$\frac{g^2}{2m_\rho^2}\int (dx)\bar{N}\gamma^\mu\tau N \cdot \partial_\mu\pi \times \pi$$

$$= \int (dx)\left(\frac{f_0}{m_\pi}\right)^2 \bar{N}\gamma^\mu\tau N \cdot \partial_\mu\pi \times \pi,$$

(3.54)

in which the connection (3.49) has been used. Notice that (3.54) is just the s-wave coupling term in (3.7).

Since the coupling constant g is known, it would seem that the interaction term $\rho_\mu \cdot \partial_\mu\pi \times \pi$ could be used to describe the instability of the ρ meson, the decay $\rho \to 2\pi$, i.e., the ρ width could be calculated. However, the ρ decay is a high-energy process, and a more elaborate high-energy theory is needed.

What is missing here is a treatment of the $\delta\varphi$ transformations analogous to the $\delta\omega$ transformations. Recall that $\delta\omega$ and $\lambda\delta\varphi$ are the six parameters associated with the group O_4. So far we have just let the parameters $\delta\omega$ be space-time dependent to introduce the ρ meson. Obviously, we should proceed analogously with the $\delta\varphi$ transformations. These considerations suggest the introduction of a 1^+, $T=1$ particle and the corresponding use of non-Abelian gauge transformations. This particle is A_1, with mass $m = 1080$ MeV. (From the experimental side, there are repeated attempts to show that it is only a kinematical effect.) Let's begin with the π field and consider its response to a $\delta\varphi(x)$ transformation,

$$\delta[(\partial_\mu + g\rho_\mu \times)\pi] = \partial_\mu \delta\varphi + g\rho_\mu \times \delta\varphi, \tag{3.55}$$

which suggests the introduction of a compensating term $-\,(1/2)m_A\mathscr{A}_\mu$, where the constant m_A is introduced for dimensional reasons. The transformation law is

$$\delta\mathscr{A}_\mu = -\,\frac{2g}{m_A}\,\delta\varphi \times \rho_\mu + \frac{2}{m_A}\,\partial_\mu \delta\varphi - \delta\omega \times \mathscr{A}_\mu, \tag{3.56}$$

where we have included the response of $\mathscr{A}_\mu$ to an isotopic rotation $\delta\omega$. Also we have

$$\delta\rho_\mu = -\delta\omega \times \rho_\mu + (1/g)\partial_\mu \delta\omega + \delta_\varphi \rho_\mu, \tag{3.57}$$

where $\delta_\varphi \rho_\mu$ is the response of ρ_μ to a chiral transformation, and is to be determined. Consider the combinations $\delta(\rho_\mu \pm \mathscr{A}_\mu)$,

$$\delta(\rho_\mu \pm \mathscr{A}_\mu) = \frac{1}{g}\partial_\mu\left(\delta\omega \pm \frac{2g}{m_A}\,\delta\varphi\right) - \left(\delta\omega \pm \frac{2g}{m_A}\,\delta\varphi\right) \times \rho_\mu$$

$$-\,\delta\omega \times (\pm\,\mathscr{A}_\mu) + \delta_\varphi \rho_\mu. \tag{3.58}$$

Since the combinations $\delta\omega \pm \left(\dfrac{2g}{m_A}\right)\delta\varphi$ appear everywhere except the last terms, the requirement of a group structure obviously demands

$$\delta_\varphi \rho_\mu = -\,\frac{2g}{m_A}\,\delta\varphi \times \mathscr{A}_\mu. \tag{3.59}$$

And we have

$$\delta(\rho_\mu \pm \mathscr{A}_\mu) = -\delta\omega_\pm \times (\rho_\mu \pm \mathscr{A}_\mu) + (1/g)\partial_\mu \delta\omega_\pm, \tag{3.60}$$

with

$$\delta\omega_\pm = \delta\omega \pm \frac{2g}{m_A}\,\delta\varphi. \tag{3.61}$$

This must be the same group as before, where we had

$$\delta\omega_{\pm} = \delta\omega \pm \frac{2f_0}{m_\pi}\delta\varphi;$$

therefore

$$\frac{g}{m_A} = \frac{f_0}{m_\pi} = \frac{g}{\sqrt{2}m_\rho},$$

and

$$m_A = \sqrt{2}m_\rho, \tag{3.62}$$

which is only a mathematical statement, not yet a physical assertion. We now introduce the quantities $\rho_{\mu\nu}$ and $\mathscr{A}_{\mu\nu}$ defined by

$$\rho_{\mu\nu} \pm \mathscr{A}_{\mu\nu} = \partial_\mu(\rho_\nu \pm \mathscr{A}_\nu) - \partial_\nu(\rho_\mu \pm \mathscr{A}_\mu)$$
$$+ g(\rho_\mu \pm \mathscr{A}_\mu) \times (\rho_\nu \pm \mathscr{A}_\nu), \tag{3.63}$$

or

$$\rho_{\mu\nu} = \partial_\mu\rho_\nu - \partial_\nu\rho_\mu + g(\rho_\mu \times \rho_\nu + \mathscr{A}_\mu \times \mathscr{A}_\nu),$$

$$\mathscr{A}_{\mu\nu} = (\partial_\mu + g\rho_\mu \times)\mathscr{A}_\nu - (\partial_\nu + g\rho_\nu \times)\mathscr{A}_\mu$$
$$= D_\mu\mathscr{A}_\nu - D_\nu\mathscr{A}_\mu. \tag{3.64}$$

A natural choice for ρ and $\mathscr{A}$ contributions to the Lagrange function $\mathscr{L}$ is

$$\mathscr{L}_{\rho\mathscr{A}} = -(1/4)[(\rho_{\mu\nu})^2 + (\mathscr{A}_{\mu\nu})^2] - (1/2)m_\rho{}^2[(\rho_\mu)^2 + (\mathscr{A}_\mu)^2]. \tag{3.65}$$

Only the combinations $(\rho + \mathscr{A})^2$ and $(\rho - \mathscr{A})^2$ can occur in $\mathscr{L}_{\rho\mathscr{A}}$ in order to satisfy the chiral invariance, yet these two structures must occur with the same weight since parity is conserved in strong interactions. The partial symmetry response of $\mathscr{L}$ is

$$\delta\mathscr{L} = -\frac{m_\rho{}^2}{g}(\rho^\mu \cdot \partial_\mu\delta\omega + \mathscr{A}^\mu \lambda \partial_\mu\delta\varphi),$$

or

$$\delta\mathscr{L} = -\frac{m_\rho{}^2}{g}\rho^\mu \cdot \partial_\mu\delta\omega - m_A\mathscr{A}^\mu \cdot \partial_\mu\delta\varphi. \tag{3.66}$$

Comparison with the low-energy behavior (3.26) indicates the necessary correspondences for low-energy behavior,

$$\rho^\mu \to \frac{g}{m_\rho{}^2}j^\mu{}_T,$$

$$\mathscr{A}^\mu \to \frac{g}{m_\rho{}^2}k^\mu. \tag{3.67}$$

If we now include the π term $\left[(\partial_\mu + g\rho_\mu \times)\pi - (1/2)m_A\mathscr{A}_\mu\right]^2$, it is immediately seen that π and $\mathscr{A}_\mu$ are mixed objects. We must carry out a diagonalization, which will also change the normalization of the π term. In anticipation of this, let us write

$$\mathscr{L}_{\pi\rho\mathscr{A}} = -a\left[D_\mu\pi - (1/2)m_A\mathscr{A}_\mu\right]^2 - (1/2)m_\pi^2\pi^2$$
$$-(1/4)\left[(\rho_{\mu\nu})^2 + (\mathscr{A}_{\mu\nu})^2\right] - (1/2)m_\rho^2(\rho_\mu^2 + \mathscr{A}_\mu^2). \qquad (3.68)$$

In order to identify the physical particle fields, it is sufficient to consider only the quadratic structure,

$$-a\left[\partial_\mu\pi - (1/2)m_A\mathscr{A}_\mu\right]^2 - (1/2)m_\pi^2\pi^2 - (1/2)m_\rho^2\mathscr{A}_\mu^2$$
$$-(1/4)(\partial_\mu\mathscr{A}_\nu - \partial_\nu\mathscr{A}_\mu)^2.$$

Now, the transformation

$$\mathscr{A}_\mu = A_{1\mu} + \frac{1}{m_A}\partial_\mu\pi, \qquad (3.69)$$

or, more generally,

$$\mathscr{A}_\mu = A_{1\mu} + \frac{1}{m_A}D_\mu\pi, \qquad (3.70)$$

gives

$$-(a/4)\left[\partial_\mu\pi - m_A A_{1\mu}\right]^2 - (1/4)\left[\partial_\mu\pi + m_A A_{1\mu}\right]^2$$
$$-(1/2)m_\pi^2\pi^2 - (1/4)(\partial_\mu A_{1\nu} - \partial_\nu A_{1\mu})^2.$$

Diagonization is achieved by the choice

$$a = 1.$$

And we get

$$\mathscr{L} = -(1/2)\left[(\partial_\mu\pi)^2 + m_\pi^2\pi^2\right] - (1/2)\left[(1/2)(\partial_\mu A_{1\nu} - \partial_\nu A_{1\mu})^2\right.$$
$$\left. + m_A^2 A_{1\mu}^2\right],$$

i.e. m_A is the mass of the A_1 particle. One compares the theoretical prediction

$$\frac{m_A}{m_\rho} = \sqrt{2} = 1.41, \qquad (3.71)$$

with the experimental value

$$\frac{m_A}{m_\rho} = \frac{1080}{760} = 1.42.$$

which are in striking agreement. The relation (3.71) was first derived by Weinberg from a different approach.

We now return to the necessary low-energy correspondence (3.67). We have already shown that ρ_μ does have the correct low-energy behavior. So we now consider its counterpart, which can be expressed in the following form:

$$m_A \mathscr{A}_\mu \sim \lambda k_\mu = \partial_\mu \pi - \frac{f}{m_\pi} \bar{N} \gamma_\mu i \gamma_5 \tau N + \dots . \tag{3.72}$$

But

$$m_A \mathscr{A}_\mu = \partial_\mu \pi + m_A A_{1\mu},$$

which shows that the π term is correct. We are left with the necessary correspondence

$$m_A A_{1\mu} \sim -\frac{f}{m_\pi} \bar{N} \gamma_\mu i \gamma_5 \tau N, \tag{3.73}$$

which anticipates a NA_1 coupling of the form

$$\mathscr{L}_{A_1 N} = -\frac{f}{m_\pi} \bar{N} \gamma^\mu i \gamma_5 \tau N \cdot m_A A_{1\mu}. \tag{3.74}$$

Let us check if this is indeed the case.

Now, the nucleon contribution

$$-\bar{N} \gamma^\mu ((1/i)\partial_\mu - g(1/2)\tau \cdot \rho_\mu)N + \text{mass term} \tag{3.75}$$

is invariant under the isotopic gauge transformation

$$\delta N = i(1/2)\tau \cdot \delta \omega N,$$

$$\delta \rho_\mu = (1/g) D_\mu \delta \omega. \tag{3.76}$$

But it is not invariant under a chiral transformation $\delta \varphi(x)$, to which the nucleon field responds as

$$\delta N = i 2 \left(\frac{g}{m_A} \right)^2 \frac{1}{2} \tau \cdot \pi \times \delta \varphi N, \tag{3.77}$$

which is a nonlinear gauge transformation depending on the π field. This suggests that one should replace ρ_μ in the $N\rho$ coupling term by a new field ρ_μ':

$$\rho_\mu' = \rho_\mu + \frac{2g}{m_A^2} \pi \times (m_A \mathscr{A}_\mu - D_\mu \pi). \tag{3.78}$$

Under a $\delta \varphi(x)$ transformation, we have

$$\delta \rho_\mu' = \frac{2g}{m_A^2} D_\mu (\pi \times \delta \varphi), \tag{3.79}$$

which, in conjunction with (3.77), correctly gives chiral invariance. In arriving at (3.79), we have neglected the variation of ρ_μ in D_μ which is outside the limited framework being considered. So far, the f term in πN coupling is missing. However, we always have the possibility of adding a phenomenological coupling which is a chiral invariant by itself. The same situation occurs also in electrodynamics at a phenomenological level. A magnetic coupling, which is gauge invariant, has to be introduced to describe the anomalous magnetic moment. In order to account for the low-energy p-wave πN scattering, one must introduce the coupling

$$\frac{f}{m_\pi} \bar{N}\gamma^\mu i\gamma_5\tau N \cdot [2D_\mu\pi - m_A\mathscr{A}_\mu], \tag{3.80}$$

which is chiral invariant if one disregards more complicated couplings. The structure $2D_\mu\pi - m_A\mathscr{A}_\mu$ is the appropriate gauge-invariant combination when $\delta\varphi$ is space-time dependent. Now the complete ρN and πN couplings are

$$-\bar{N}\gamma^\mu\left(\frac{1}{i}\partial_\mu - g\frac{1}{2}\tau\cdot\rho_\mu'\right)N + \frac{f}{m_\pi}\bar{N}\gamma^\mu i\gamma_5\tau N \cdot (D_\mu\pi - m_A A_{1\mu}), \tag{3.81}$$

which indeed predicts a NA_1 coupling of the form (3.74) demanded by consistency requirement.

If we turn to high-energy processes, the vector current will appear through the intermediary of the ρ meson, and the axial-vector current will appear through the intermediary of the A_1 particle. Thus we learn that nucleon vector and axial-vector form factors are governed by the ρ and A_1 masses, respectively.

3.4   Widths of ρ and A_1

We can now use our results to predict the observed properties of ρ and A_1, in particular the ρ width and A_1 width. The Lagrange function describing π, ρ, and A_1 is

$$\mathscr{L} = -(D_\mu \pi - (1/2)m_A \mathscr{A}_\mu)^2 - (1/2)m_\pi^2 \pi^2$$
$$-(1/4)\left[(\rho_{\mu\nu})^2 + (\mathscr{A}_{\mu\nu})^2\right] - (1/2)m_\rho^2\left[(\rho_\mu)^2 + (\mathscr{A}_\mu)^2\right], \quad (3.82)$$

with

$$D_\mu = \partial_\mu + g\rho_\mu \times ,$$

$$\rho_{\mu\nu} = \partial_\mu \rho_\nu - \partial_\nu \rho_\mu + g(\rho_\mu \times \rho_\nu + \mathscr{A}_\mu \times \mathscr{A}_\nu),$$

$$\mathscr{A}_{\mu\nu} = D_\mu \mathscr{A}_\nu - D_\nu \mathscr{A}_\mu. \quad (3.83)$$

When we introduce

$$\mathscr{A}_\mu = A_{1\mu} + \frac{1}{m_A} D_\mu \pi,$$

this becomes

$$\mathscr{L} = -(1/2)\left[(D_\mu \pi)^2 + m_\pi^2 \pi^2\right] - (1/4)(\rho_{\mu\nu})^2 - (1/2)m_\rho^2(\rho_\mu)^2$$
$$-(1/4)(\mathscr{A}_{\mu\nu})^2 - (1/2)m_A^2(A_{1\mu})^2, \quad (3.84)$$

where

$$\rho_{\mu\nu} = \partial_\mu \rho_\nu - \partial_\nu \rho_\mu + g(\rho_\mu \times \rho_\nu + A_{1\mu} \times A_{1\nu})$$

$$+ \frac{g}{m_A}(A_{1\mu} \times D_\nu \pi - A_{1\nu} \times D_\mu \pi) + \frac{g}{m_A^2} D_\mu \pi \times D_\nu \pi, \quad (3.85)$$

and

$$\mathscr{A}_{\mu\nu} = D_\mu A_{1\nu} - D_\nu A_{1\mu} + \frac{g}{m_A} \rho_{\mu\nu} \times \pi, \quad (3.86)$$

which is obtained with the aid of the relation

$$[D_\mu, D_\nu] = g\rho_{\mu\nu} \times . \quad (3.87)$$

We now pick out the $\rho\pi\pi$ and $A_1 \rho\pi$ coupling terms. These are

$$\mathscr{L}_{\rho\pi\pi} = g\rho^\mu \cdot \partial_\mu \pi \times \pi - \frac{1}{4}\frac{g}{m_\rho^2} \rho^{\mu\nu} \cdot \partial_\mu \pi \times \partial_\nu \pi,$$

$$\mathscr{L}_{A_1\rho\pi} = -\frac{1}{2}\frac{g}{m_A} \rho^{\mu\nu} \cdot (A_{1\mu} \times \partial_\nu \pi - A_{1\nu} \times \partial_\mu \pi)$$

$$-\frac{g}{2m_A} A_{1\mu\nu} \cdot \rho_{\mu\nu} \times \pi. \quad (3.88)$$

These expressions can be simplified,

$$\mathcal{L}_{\rho\pi\pi} = g\rho^\mu \cdot \partial_\mu\pi \times \pi - \frac{1}{4}\frac{g}{m_\rho^2}\rho^{\mu\nu} \cdot \partial_\nu(\partial_\mu\pi \times \pi),$$

$$\mathcal{L}_{A_1\rho\pi} = \frac{g}{m_A}\rho^{\mu\nu} \cdot \partial_\mu(A_{1\nu} \times \pi). \tag{3.89}$$

When these couplings are applied to real ρ particles, further simplifications can be introduced through the use of the field equation

$$\partial_\mu\rho^{\mu\nu} = m_\rho^2\rho^\nu,$$

Thus,

$$\mathcal{L}_{\rho\pi\pi} \rightarrow (3/4)g\rho^\mu \cdot \partial_\mu\pi \times \pi, \tag{3.90}$$

$$\mathcal{L}_{A_1\rho\pi} \rightarrow -(1/2)gm_A\rho^\mu \cdot A_{1\mu} \times \pi. \tag{3.91}$$

Note the factor (3/4) in the $\rho\pi\pi$ coupling. A straightforward calculation gives the widths for ρ and A_1:

$$\Gamma_\rho = \frac{3}{8}\frac{g^2}{4\pi}\frac{p^3}{m_\rho^2}, \tag{3.92}$$

$$\Gamma_{A_1} = \frac{1}{4}\frac{g^2}{4\pi}\left[1 + \frac{1}{3}\left(\frac{p}{m_\rho}\right)^2\right]p, \tag{3.93}$$

where p is the momentum of the decay products in the center-of-mass frame. For $\rho \rightarrow 2\pi$, $p = 350$ MeV, and for $A_1 \rightarrow \rho\pi$, $p = 245$ MeV. The predicted values are

$$\Gamma_\rho \simeq 100 \text{ MeV} \tag{3.94}$$

[without the factor (3/4), the result would be $\Gamma_\rho \sim 180$ MeV] and

$$\Gamma_{A_1} \simeq 200 \text{ MeV}. \tag{3.95}$$

Experimentally, the average width of ρ is 125 MeV (116 MeV for ρ^0 and 132 MeV for $\rho^\pm$), and the width of A_1 is 130 ± 40 MeV. The theoretical results are certainly qualitatively right, but indicate that the predicted Γ_ρ is too small and Γ_{A_1} is too big. However, the qualitative consistency lends support to the underlying theoretical ideas.

Of course, we have used only the simplest possible theory in which no adjustable parameter appears. As in the discussion of πN coupling, it is always possible to add individually invariant terms with arbitrary coefficients. A relevant example here is

$$\gamma g(1/2)\rho^{\mu\nu\prime} \cdot \mathcal{A}_\mu' \times \mathcal{A}_\nu', \tag{3.96}$$

where

$$\mathcal{A}_\mu' = \mathcal{A}_\mu - \frac{2}{m_A} D_\mu \pi = A_{1\mu} - \frac{1}{m_A} D_\mu \pi, \tag{3.97}$$

and

$$\rho_\mu' \simeq \rho_\mu, \tag{3.98}$$

are chiral invariant combinations as far as our problem is concerned. The contributions to $\rho\pi\pi$ and $A_1\rho\pi$ couplings are

$$\frac{1}{2} \gamma \frac{g}{m_A^2} \rho^{\mu\nu} \cdot \partial_\mu \pi \times \partial_\nu \pi - \frac{\gamma g}{m_A} \rho^{\mu\nu} \cdot A_{1\mu} \times \partial_\nu \pi. \tag{3.99}$$

The first term corrects the effective $\rho\pi\pi$ coupling constant,

$$\rho\pi\pi: \quad (3/4)g \rightarrow ((3/4) + (1/4)\gamma)g. \tag{3.100}$$

We now consider an approximate evaluation of the $A_1\rho\pi$ additional coupling, based on the nonrelativistic behavior of the ρ meson in the decay $A_1 \rightarrow \rho + \pi$ $[(p/m_\rho)^2 \sim (1/10)]$. For a particular decay, $A_1{}^0 = 0$ in the rest frame of A_1, and

$$\rho^{\mu\nu} \cdot A_{1\mu} \times \partial_\nu \pi \rightarrow -(p_\rho{}^k \rho^\nu - p_\rho{}^\nu \rho^k) \cdot A_{1k} \times p_{\pi\nu} \pi. \tag{3.101}$$

Here

$$p_{\pi\nu} \rho^\nu = (P_\nu - p_{\rho\nu}) \rho^\nu = -m_A \rho^0, \tag{3.102}$$

where P_ν is the total energy-momentum in the decay process. But

$$\rho^0 \simeq \frac{\mathbf{p} \cdot \boldsymbol{\rho}}{m_\rho} \tag{3.103}$$

and therefore

$$p_{\pi\nu} \rho^\nu \simeq -m_A \frac{\mathbf{p} \cdot \boldsymbol{\rho}}{m_\rho} \tag{3.104}$$

Also

$$-p_\rho p_\pi = -(1/2)(P^2 - p_\rho{}^2 - p_\pi{}^2)$$
$$= (1/2)(m_A{}^2 - m_\rho{}^2 - m_\pi{}^2),$$

or

$$-p_\rho p_\pi \simeq (1/2)m_\rho{}^2, \tag{3.105}$$

neglecting the pion mass. Thus

$$\rho^{\mu\nu} \cdot A_{1\mu} \times \partial_\nu \pi \simeq -\left(\frac{m_\rho{}^2}{2} \rho^k - m_A \frac{p^k \mathbf{p} \cdot \boldsymbol{\rho}}{m_\rho}\right) \cdot A_{1k} \times \pi,$$

or

$$\rho^{\mu\nu} \cdot A_{1\mu} \times \partial_\nu \pi \simeq -\frac{m_\rho^2}{2} \rho^k \cdot A_{1k} \times \pi$$

$$= -\frac{m_A^2}{4} \rho^\mu \cdot A_{1\mu} \times \pi. \tag{3.106}$$

This gives the additional $A_1 \rho \pi$ coupling

$$+\gamma(1/4)gm_A\rho^\mu \cdot A_{1\mu} \times \pi, \tag{3.107}$$

and the effective $A_1 \rho \pi$ coupling constant becomes

$$A_1\rho\pi: \ g \rightarrow g(1-(1/2)\gamma). \tag{3.108}$$

If γ is chosen to be positive, it will increase Γ_ρ and decrease Γ_{A_1}. The choice

$$\gamma = (2/5) \tag{3.109}$$

predicts

$$\Gamma_\rho \ = 125 \ \text{MeV},$$

$$\Gamma_{A_1} = 130 \ \text{MeV}, \tag{3.110}$$

in complete accordance with the experimental numbers.

3.5   Low-Energy $\pi\pi$ Interactions

Our theory of π, N, and ρ chiral transformations considered so far is mathematically (but not physically) inconsistent. The nucleon chiral transformation, for example,

$$\delta N = i \left(\frac{g}{m_A}\right)^2 \tau \cdot \pi \times \delta\varphi N, \tag{3.111}$$

is non-Abelian, while the π transformation

$$\delta\pi = \delta\varphi, \tag{3.112}$$

is Abelian. This is because we have not considered $\pi\pi$ interactions. Recall that the nucleon interaction properties were used to identify the chiral group structure. If one considers a π system, in which the nucleon does not appear, the physical evidence of the non-Abelian chiral group can only be revealed through the interactions of the π mesons. We now extend the π transformation law and introduce $\pi\pi$ interactions by insisting on the group structure established for the $\pi+N$ system under limited physical circumstances. Since the transformation (3.112) through the ρ coupling induces an isotopic rotation with $\delta\omega = 2(g/m_A)^2\pi \times \delta\varphi$, any isotopic-spin bearing field must respond universally. Thus,

$$\delta\pi = -2\left(\frac{g}{m_A}\right)^2 (\pi \times \delta\varphi) \times \pi + \delta\varphi \left[1 + \left(\frac{g}{m_A}\pi\right)^2\right], \tag{3.113}$$

in which we anticipate that a scale dilation of $\delta\varphi$ is also required for the complete group property. The group property for the transformation law (3.113) can be checked, as before, by comparing successive transformations in one order

$$\pi'' = \pi' + \delta_2\varphi \left[1 + \left(\frac{g}{m_A}\pi'\right)^2\right] - 2\left(\frac{g}{m_A}\right)^2 (\pi' \times \delta_2\varphi) \times \pi',$$

$$\pi' = \pi + \delta_1\varphi \left[1 + \left(\frac{g}{m_A}\pi\right)^2\right] - 2\left(\frac{g}{m_A}\right)^2 (\pi \times \delta_1\varphi) \times \pi, \tag{3.114}$$

and in opposite order. To illustrate the procedure, consider the simplest terms, up to cubic in π and $\delta_1\varphi$, $\delta_2\varphi$.

$$\pi'' = \pi + \delta_1\varphi \left[1 + \left(\frac{g}{m_A}\pi\right)^2\right] + \delta_2\varphi \left[1 + \left(\frac{g}{m_A}\pi\right)^2\right]$$

$$+ 2\left(\frac{g}{m_A}\right)^2 \delta_2\varphi\pi \cdot \delta_1\varphi - 2\left(\frac{g}{m_A}\right)^2 (\pi \times \delta_2\varphi) \times \delta_1\varphi$$

$$- 2\left(\frac{g}{m_A}\right)^2 (\delta_1\varphi \times \delta_2\varphi) \times \pi, \tag{3.115}$$

or

$$\pi'' = \pi - 2\left(\frac{g}{m_A}\right)^2 (\delta_1\varphi \times \delta_2\varphi) \times \pi$$
$$+ (\delta_1\varphi + \delta_2\varphi)\left[1 + \left(\frac{g}{m_A}\right)^2 \pi^2\right]$$
$$+ 2\left(\frac{g}{m_A}\right)^2 \pi\delta_1\varphi \cdot \delta_2\varphi. \tag{3.116}$$

Upon comparing with the transformations in reversed order, all terms symmetrical in $\delta_1\varphi$ and $\delta_2\varphi$ disappear, and we get *exactly*

$$\delta_{[12]}\pi = -\delta_{[12]}\omega \times \pi, \tag{3.117}$$

with

$$\delta_{[12]}\omega = \left(\frac{2g}{m_A}\right)^2 \delta_1\varphi \times \delta_2\varphi, \tag{3.118}$$

since all higher order terms cancel completely. The nucleon transformation law (3.111) is now also exact. The π transformation law (3.113) can also be exhibited in the following form

$$\delta\pi = \delta\varphi + \left(\frac{g}{m_A}\right)^2 \left[2\pi\delta\varphi \cdot \pi - \delta\varphi\pi^2\right]. \tag{3.119}$$

We have exhibited a mathematical object which is transformed very simply according to (3.119). But it is not necessarily *the* π field. It is always possible to redefine the π field in such a way that the original physical identification in terms of noninteracting particles is not changed. Nevertheless, the redefinition may change interactions. Only experiment can determine which is the correct identification. Let's recognize this ambiguity, and write generally, in an infinite series

$$\pi \rightarrow \pi\left[1 + (1-\beta)\left(\frac{g}{m_A}\pi\right)^2 + \dots\right]. \tag{3.120}$$

This general π field has the transformation property

$$\delta\pi = \delta\varphi\left[1 - (2-\beta)\left(\frac{g}{m_A}\pi\right)^2\right] + 2\beta\left(\frac{g}{m_A}\right)^2 \pi\delta\varphi \cdot \pi + \dots. \tag{3.121}$$

We now return to $\mathscr{L}_\pi$ and ask for its chiral response. It is

$$\delta\mathscr{L}_\pi = -\partial^\mu\pi\partial_\mu\delta\pi - m_\pi{}^2\pi \cdot \delta\pi. \tag{3.122}$$

Substitution of (3.121) for $\delta\pi$ gives

$$\delta\mathscr{L}_\pi + m_\pi^2\pi\cdot\delta\varphi = \delta\left[\frac{1}{4}(2-\beta)\left(\frac{g}{m_A}\right)^2(\partial_\mu\pi^2)^2 - \beta\left(\frac{g}{m_A}\right)^2\right.$$

$$\left.\times(\partial_\mu\pi)^2\pi^2 - \beta\left(\frac{g}{m_A}\right)^2\frac{1}{4}(\partial_\mu\pi^2)^2 + \frac{1}{4}m_\pi^2(2-3\beta)(\pi^2)^2\right] + \dots . \quad (3.123)$$

It is important to recognize that we have the option of insisting that

$$\delta\mathscr{L} + m_\pi^2\pi\cdot\delta\varphi = 0 \qquad\qquad (3.124)$$

still holds, if necessary by a redefinition of π produced by a suitable choice of β, etc. This is a convention for the π field, not a hypothesis. Then the combination

$$\mathscr{L} = \mathscr{L}_\pi + \mathscr{L}_{\pi\pi} + \dots$$

with

$$\mathscr{L}_{\pi\pi} = -\frac{1}{2}\left(\frac{g}{m_A}\right)^2\left[(1-\beta)(\partial_\mu\pi^2)^2 - 2\beta(\partial_\mu\pi)^2\pi^2\right.$$

$$\left. + m_\pi^2\left(1-\frac{3}{2}\beta\right)(\pi^2)^2\right], \qquad (3.125)$$

is partially chiral invariant, broken only by the pion mass term in accordance with (3.124). For the application to π-π scattering, the π field satisfies the equation

$$(-\partial^2 + m_\pi^2)\pi = 0, \qquad\qquad (3.126)$$

and

$$(\partial_\mu\pi^2)^2 + 2(\partial_\mu\pi)^2\pi^2 = 2\partial_\mu\pi\cdot\partial_\mu(\pi\pi^2)$$

$$\to -2\partial^2\pi\cdot\pi\pi^2$$

$$= -2m_\pi^2(\pi^2)^2. \qquad (3.127)$$

Consequently,

$$\mathscr{L}_{\pi\pi} \to -\frac{1}{2}\left(\frac{g}{m_A}\right)^2\left[(\partial_\mu\pi^2)^2 + m_\pi^2\left(1+\frac{1}{2}\beta\right)(\pi^2)^2\right],$$

$$\left(\frac{g}{m_A} = \frac{f_0}{m_\pi}\right). \qquad (3.128)$$

All sensitivity to the model parameter β is in the pion mass term.

The physical background of these considerations should be kept in mind. Chiral invariance is recognized from the strong-interaction phenomenology of the low-energy $\pi + N$ system. The conclusions are therefore mainly about low-energy processes. The possibility of generalizing to high-energy processes is of course a tempting one, and in fact we have done so in the discussion of $A_1\rho\pi$ dynamics. But as a local interaction, (3.128) is expected to be only applicable in the low-energy region. Thus we apply it to the π-π scattering at the threshold. The implied scattering amplitudes for $T = 0, 2$ states are

$$m_\pi a_0 = \frac{f_0{}^2}{4\pi}\left[6 - \frac{5}{2}\left(1 + \frac{1}{2}\beta\right)\right],$$

$$m_\pi a_2 = \frac{f_0{}^2}{4\pi}\left[-\left(1 + \frac{1}{2}\beta\right)\right]. \tag{3.129}$$

A model-independent combination is

$$m_\pi\left(a_0 - \frac{5}{2}a_2\right) = 6\frac{f_0{}^2}{4\pi} = 0.34 \pm 0.02. \tag{3.130}$$

However, no direct experimental comparison is available.

3.6   Spectrum of the Decay Process $\eta^*(960) \rightarrow \eta + 2\pi$

There is a closely related strong-interaction process for which some experimental information is available. It is the decay

$$\eta^*(960) \rightarrow \eta + 2\pi, \tag{3.131}$$

and the accompanying deviation of the η energy distribution from purely phase-space considerations. This is at the moment the only experimentally known strong-interaction process which involves just spin-0 particles. Both η and η^* are $T = 0, 0^-$ particles.

We must first consider the unitary generalization of our chiral transformation. Let ϕ be a $n \times n$ Hermitian matrix and consider the infinitesimal unitary transformation ($\delta\omega$ is an infinitesimal Hermitian matrix),

$$\delta\phi = i[\delta\omega, \phi]. \tag{3.132}$$

The group property is

$$\delta_{[12]}\omega = (1/i)[\delta_1\omega, \delta_2\omega]. \tag{3.133}$$

For an SU_2 matrix, we have

$$\phi = (1/2)\tau \cdot \pi, \tag{3.134}$$

and

$$\delta\omega = (1/2)\tau \cdot \delta\omega, \tag{3.135}$$

where the $\delta\omega$ on the right-hand side is a vector, while the $\delta\omega$ on the left-hand side is a matrix. Now (3.132) simply reduces to the ordinary isotopic rotation of the π field.

The general chiral transformation on ϕ is

$$\delta\phi = \delta\varphi + \lambda^2 \phi\delta\varphi\phi. \tag{3.136}$$

The exact group property

$$\delta_{[12]}\phi = i[\delta_{[12]}\omega, \phi], \tag{3.137}$$

with

$$\delta_{[12]}\omega = (1/i)[\lambda\delta_1\varphi, \lambda\delta_2\varphi], \tag{3.138}$$

can be readily verified. For SU_2, $\delta\varphi = (1/2)\tau \cdot \delta\varphi$, and (3.136) gives the previous chiral transformation of the π field, since

$$\tau \cdot \pi \, \tau \cdot \delta\varphi \, \tau \cdot \pi = \tau \cdot [2\pi\delta\varphi \cdot \pi - \delta\varphi\pi^2]. \tag{3.139}$$

The complete group structure of the transformations (3.132) and (3.136) is $U_n \times U_n$ if the parameters $\delta\omega \pm \lambda\delta\varphi$ are introduced.

For the present purpose, we are interested in the generalization of SU_2 to U_2, and the field matrix ϕ is

$$\phi = (1/2)(\eta_2 + \tau \cdot \pi), \tag{3.140}$$

where

$$\eta_2 = \cos \vartheta \eta + \sin \vartheta \eta^*, \tag{3.141}$$

is the projection of a full U_3 object into the U_2 subspace, and is a mixture of the physical particles η and η^*. There will be another combination η_3, orthogonal to η_2, to represent the projection along the third axis. We shall not consider chiral transformations associated with η_2. The π chiral transformation is now represented by

$$\delta\varphi = (1/2)\tau \cdot \delta\varphi, \tag{3.142}$$

and (3.136) implies the differential transformation laws for π and η_2,

$$\delta\pi = \delta\varphi + ((1/2)\lambda)^2 \left[2\pi\delta\varphi \cdot \pi + \delta\varphi(\eta_2{}^2 - \pi^2) \right],$$

$$\delta\eta_2 = 2((1/2)\lambda)^2 \delta\varphi \cdot \pi\eta_2. \tag{3.143}$$

Incidentally, the combination $\eta_2 / \left[1 + ((1/2)\lambda\pi)^2 \right]$ is a chiral invariant,

$$\delta \left[\frac{\eta_2}{1 + \left(\dfrac{1}{2}\lambda\pi \right)^2} \right] = 0, \tag{3.144}$$

if terms cubic in η_2 are ignored. Again let's recognize the possibility of redefining fields and write more generally

$$\pi \to \pi \left[1 + (1-\beta)((1/2)\lambda\pi)^2 - (1-\beta_1)((1/2)\lambda\eta_2)^2 + \dots \right],$$

$$\eta_2 \to \eta_2 \left[1 + (1-\beta_2)((1/2)\lambda\pi)^2 \right.$$

$$\left. - (1-\beta_3)((1/2)\lambda\eta_2)^2 + \dots \right].$$

Then

$$\delta\pi = \delta\varphi \left[1 - (2-\beta)((1/2)\lambda\pi)^2 + (2-\beta_1)((1/2)\lambda\eta_2)^2 \right]$$

$$+ (1/2)\beta\lambda^2 \pi\delta\varphi \cdot \pi,$$

$$\delta\eta_2 = (1/2)\beta_2\lambda^2 \delta\varphi \cdot \pi\eta_2. \tag{3.145}$$

We now return to $\mathscr{L}_\pi$ and investigate the η_2-dependent part of its response (the η_2-independent part will simply lead to $\mathscr{L}_{\pi\pi}$ which we have already discussed). It is

$$(\delta\mathscr{L}_\pi)_{\eta_2} = -\partial^\mu\pi \cdot \delta\varphi(2-\beta_1)((1/2)\lambda)^2 \partial_\mu\eta_2{}^2$$

$$- m_\pi{}^2 \pi \cdot \delta\varphi(2-\beta_1)((1/2)\lambda\eta_2)^2.$$

The additional coupling which will provide a compensating term is therefore

$$\mathcal{L}_{\pi\eta_2} = \left(\frac{f_0}{m_\pi}\right)^2 \left(1 - \frac{1}{2}\beta_1\right)\left[\partial^\mu\pi^2\partial_\mu\eta_2^{\,2} + m_\pi^{\,2}\pi^2\eta_2^{\,2}\right], \tag{3.146}$$

in which η_2 is treated as a chiral invariant. Now (3.146) implies the $\eta^*\eta\pi$ coupling

$$\mathcal{L}_{\eta^*\eta\pi} = \left(\frac{f_0}{m_\pi}\right)^2 \left(1 - \frac{1}{2}\beta_1\right)\sin 2\vartheta\left[\partial^\mu\pi^2\partial_\mu(\eta\eta^*)\right.$$

$$\left. + m_\pi^{\,2}\pi^2\eta\eta^*\right], \tag{3.147}$$

using (3.141). It is interesting to notice that all the unknown parameters ϑ and β_1 are lumped together as a multiplicative factor.

Analogous considerations on $\mathcal{L}_\eta + \mathcal{L}_{\eta^*}$ will give new couplings, but they do not contribute to the decay process $\eta^* \to \eta + 2\pi$. This can be seen from the following. From (3.144) and a similar statement about η_3 it follows that substitutions of the form

$$\eta \to \eta + a\eta^*\pi^2,$$

$$\eta^* \to \eta^* + b\eta\pi^2,$$

will make $\mathcal{L}_\eta + \mathcal{L}_{\eta^*}$ chiral invariant as far as the present application is concerned. The additional terms so obtained which contribute to the decay $\eta^* \to \eta + 2\pi$ are

$$-\partial_\mu\eta\partial^\mu(a\eta^*\pi^2) - m_\eta^{\,2}\eta a\eta^*\pi^2$$

$$-\partial_\mu\eta^*\partial^\mu(b\eta\pi^2) - m_{\eta^*}^{\,2}\eta^* b\eta\pi^2$$

$$\to (\partial^2 - m_\eta^{\,2})\eta \cdot a\eta^*\pi^2 + (\partial^2 - m_{\eta^*}^{\,2})\eta^* \cdot b\eta\pi^2 = 0, \tag{3.148}$$

since η and η^* are being used as fields of real particles. Therefore, there is no contribution from $\mathcal{L}_\eta + \mathcal{L}_{\eta^*}$. This is a consequence of the stationary action principle.

The structureless coupling $\pi^2\eta\eta^*$ simply introduces an invariant phase-space factor. For the decay

$$\eta^*(P) \to \eta(p) + 2\pi, \tag{3.149}$$

as described by (3.147) there is an additional factor

$$-(P-p)^2 - m_\pi^{\,2} = m_{\eta^*}^{\,2} + m_\eta^{\,2} - m_\pi^{\,2} - 2m_{\eta^*} + (m_\eta + T_\eta)$$

$$= (m_{\eta^*} - m_\eta)^2 - m_\pi^{\,2} - 2m_{\eta^*}T_\eta, \tag{3.150}$$

where T_η is the kinetic energy of η. It is very useful to note that

$$m_{\eta^*} = 7m_\pi,$$

$$m_\eta = 4m_\pi, \tag{3.151}$$

within experimental error. Experimental analysis is expressed in terms of the Dalitz parameter

$$y = \frac{m_\eta + 2m_\pi}{m_\pi} \frac{T_\eta}{m_{\eta^*} - m_\eta - 2m_\pi} - 1, \quad -1 < y < \frac{8}{7}, \tag{3.152}$$

and (3.150) becomes

$$1 - \frac{7}{17} y, \tag{3.153}$$

apart from a multiplicative factor. Thus, the spectrum will deviate from the pure phase-space prediction by an asymmetry factor

$$\left(1 - \frac{7}{17} y\right)^2. \tag{3.154}$$

Comparison with experiment (102 events $-$ 21 background) is listed as follows.

	$y = -0.8$,	-0.4,	0,	0.4,	0.8,
Experiment	26 ± 8,	17 ± 5,	15 ± 5,	11 ± 3,	12 ± 4,
$15\frac{1}{3}\left(1 - \frac{7}{17} y\right)^2$	27,	21,	15,	11,	7,

The normalization is adjusted to the total number of events after removing the background. The theoretical predictions are in reasonable agreement with experiment.

3.7   Pion Electromagnetic Mass

We shall now discuss the superposition of electromagnetic effects on strong interactions through the requirement of gauge invariance. Let's begin with the isotopic-spin non-Abelian gauge invariance, broken only by the ρ mass term. The ρ transformation and the response of the Lagrange function are

$$\delta\rho_\mu = -\delta\omega \times \rho_\mu + \frac{1}{g}\partial_\mu\delta\omega,$$

$$\delta\mathscr{L} = -\frac{m_\rho^2}{g}\rho^\mu \cdot \partial_\mu\delta\omega. \tag{3.155}$$

The latter implies, incidentally, that

$$\partial_\mu\rho^\mu = 0.$$

Suppose we pick out one axis, say the third axis, and consider $\delta\omega$ directed along this axis; we find

$$\delta g\rho_{\mu3} = \partial_\mu\delta\omega, \tag{3.156}$$

which is an Abelian gauge transformation. We now recognize the possibility of maintaining invariance under this transformation through the compensating effect of the electromagnetic gauge transformation

$$\delta eA_\mu = \partial_\mu\delta\omega, \tag{3.157}$$

and we have

$$\delta(g\rho_{\mu3} - eA_\mu) = 0. \tag{3.158}$$

Thus we propose to incorporate the electromagnetic effects by the generalization of the ρ mass term to

$$-\frac{1}{2}m_\rho^2(\rho_{\mu1,2})^2 - \frac{1}{2}m_\rho^2\left(\rho_{\mu3} - \frac{e}{g}A_\mu\right)^2. \tag{3.159}$$

Gauge invariance now appears as a combined property. When $\delta\omega$ is directed along the third axis, $\rho_{\mu1}$ and $\rho_{\mu2}$ rotate in a way characteristic of charged fields, leaving the first term in (3.159) unchanged, while the inhomogeneous transformation of $\rho_{\mu3}$, (3.156), is compensated by the gauge transformation of A_μ, (3.157). One must verify that the constant e in (3.159) is indeed the physical charge. For this, imagine a strongly interacting system is subject to an external potential, which acts as a driving term. The major effect is to displace the ρ field,

$$\rho_{\mu3} \rightarrow \rho_{\mu3} + \frac{e}{g}A_\mu. \tag{3.160}$$

Thus, the nucleon coupling, for example, becomes

$$g\bar{N}\gamma^{\mu}(1/2)\tau N \cdot \rho_{\mu} \to g\bar{N}\gamma^{\mu}(1/2)\tau N \cdot \rho_{\mu}$$
$$+ \bar{N}\gamma^{\mu}e(1/2)\tau_3 A_{\mu}N, \tag{3.161}$$

which reproduces correctly the isotopic-spin-dependent part of electrical charge. The isotopic-spin scalar part is obtained by a similar treatment for the other neutral 1^- particles, ω and ϕ.

The coupling between the photon and strongly interacting particles is

$$m_{\rho}^{2}\frac{e}{g}\rho_3^{\mu}A_{\mu}. \tag{3.162}$$

Through the identification of a photon source, (3.162) implies a coupling in the action, through photon exchange,

$$w_e = \frac{1}{2}\left(\frac{m_{\rho}^{2}e}{g}\right)^2 \int (dx)(dx')\rho_3^{\mu}(x)D_+(x-x')\rho_{\mu3}(x'). \tag{3.163}$$

Thus, the introduction of the photon modifies the strong-interaction phenomenology. Among other things, it produces a mass splitting between electrically neutral and charged particles. As a simplest example, we apply (3.163) to the pion electromagnetic mass splitting by using chiral transformations. The idea is as follows. The pion mass term breaks chiral symmetry. Now the action w_e will produce an additional violation which can be identified as a change of the pion mass term. Under a chiral transformation we have

$$\delta\rho_{\mu} = -\frac{2g}{m_A}\delta\varphi \times \mathscr{A}_{\mu},$$

$$\delta\mathscr{A}_{\mu} = -\frac{2g}{m_A}\delta\varphi \times \rho_{\mu}, \tag{3.164}$$

where $\mathscr{A}_{\mu}$ is the axial-vector gauge field associated with chiral transformations.

Since chiral symmetry is broken, the divergenceless condition on ρ^{μ} is violated under a chiral transformation, (3.164). But, as a photon source, it must be conserved. In order to guarantee that only the conserved part contributes, we insert a projection operator in (3.163), i.e.,

$$\rho_3^{\mu}(x)D_+(x-x')\rho_{3\mu}(x') \to \rho_3^{\mu}(x)D_{\mu\nu}(x-x')\rho_3^{\nu}(x'), \tag{3.165}$$

where

$$\partial_{\mu}D^{\mu\nu} = 0. \tag{3.166}$$

In momentum space, $D_{\mu\nu}$ has the form

$$D_{\mu\nu}(p) = \frac{g_{\mu\nu} - \dfrac{p_\mu p_\nu}{p^2}}{p^2 - i\varepsilon}. \tag{3.167}$$

Now, the chiral response of the action w_e is

$$\delta_\varphi w_e = \left(\frac{m_\rho^2 e}{g}\right)^2 \frac{2g}{m_A} \int (dx)(dx')$$
$$\times (\mathscr{A}^\mu \times \delta\varphi)_3(x) D_{\mu\nu}(x-x')\rho_3{}^\nu(x'), \tag{3.168}$$

and

$$\delta_\varphi{}^2 w_e = \left(\frac{m_\rho^2 e}{g}\right)^2 \left(\frac{2g}{m_A}\right)^2 \int (dx)(dx') D_{\mu\nu}(x-x')$$
$$\times \{(\mathscr{A}^\mu \times \delta\varphi)_3(x) \cdot (\mathscr{A}^\nu \times \delta\varphi)_3(x')$$
$$+ [(\rho^\mu \times \delta\varphi) \times \delta\varphi]_3(x)\rho_3{}^\nu(x')\}. \tag{3.169}$$

Let us emphasize that we are now considering processes that are not included in the strong-interaction phenomenology. A typical term in (3.169) contains two fields. $\mathscr{A}^\mu(x)\mathscr{A}^\nu(x')$, or $\rho^\mu(x)\rho^\nu(x')$, which operate at two different

FIGURE 3.1

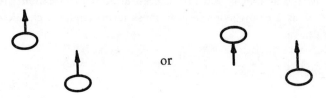

or

FIGURE 3.2

space-time points, and describe the emission or absorption of a particle, as shown in Fig. 3.1. Included among all these processes is the possibility that one of the two sources absorbs the particle emitted by the other (Fig. 3.2). That is, any two-particle term also includes the virtual exchange of one .particle. All this is contained in the general vacuum amplitude for noninteracting particles

$$\langle 0_+ | 0_- \rangle^S = \exp\left[(i/2)\int (dx)(dx')S(x)G_+(x-x')S(x')\right]. \tag{3.170}$$

Let

$$S \to S_1 + S_2 + S,$$

and note that those terms which depend on S_1 and S_2 are

$$\exp\left[i\int (dx)(dx')S_1(x)G_+(x-x')S_2(x') + i\int (dx)S_1(x)\chi(x)\right.$$
$$\left. + i\int (dx)\chi(x)S_2(x)\right], \tag{3.171}$$

where the field $\chi(x)$ is

$$\chi(x) = \int (dx')G_+(x-x')S(x'). \tag{3.172}$$

For weak sources, the terms bilinear in S_1 and S_2 are

$$-\int (dx)(dx')S_1(x)\left[\chi(x)\chi(x') - iG_+(x-x')\right]S_2(x'). \tag{3.173}$$

Accordingly, the virtual exchange process is produced from the process involving two real particles by the substitution

$$\chi(x)\chi(x') \to -iG_+(x-x'). \tag{3.174}$$

Isotopic-spin invariance demands

$$G_+(x-x')_{ab} = \delta_{ab}G_+(x-x'), \; a, b = 1, 2, 3, \tag{3.175}$$

where a and b are isotopic-spin indices. The substitution (3.174) gives

$$\delta_\varphi^2 w_e = 2e^2 m_\rho^2 i\int (dx)(dx')D_{\mu\nu}(x-x')\left[G^{\mu\nu}(x-x')_\rho\right.$$
$$\left. - G^{\mu\nu}(x-x')_a\right](\delta\varphi_{ch})^2, \tag{3.176}$$

where $\delta\varphi_{ch}$ refers to charged components. In momentum-space representation, we have

$$\delta_\varphi^2 w_e = -2e^2 m_\rho^2 \int (dx)(\delta\varphi_{ch})^2 \frac{1}{i}\int \frac{(dp)}{(2\pi)^4} D^{\mu\nu}(p)\left[G_{\mu\nu}(p)_\rho\right.$$
$$\left. - G_{\mu\nu}(p)_a\right]. \tag{3.177}$$

This is identified as the chiral response of the additional charged pion mass term

$$-(1/2)\int (dx)\delta m_\pi^2(\pi_{ch})^2, \tag{3.178}$$

and gives

$$\delta m_\pi^2 = m_{\pi\pm}^2 - m_{\pi^0}^2$$
$$= 2e^2 m_\rho^2 \frac{1}{i}\int \frac{(dp)}{(2\pi)^4} D^{\mu\nu}(p)\left[G_{\mu\nu}(p)_\rho - G_{\mu\nu}(p)_a\right]. \tag{3.179}$$

Now

$$G_{\mu\nu}(p)_\rho = \frac{g_{\mu\nu} + \dfrac{p_\mu p_\nu}{m_\rho^2}}{p^2 + m_\rho^2 - i\varepsilon},$$

(3.180)

and since

$$\mathscr{A}_\mu = A_{1\mu} + \frac{1}{m_A} \partial_\mu \pi,$$

(3.181)

we also have

$$G_{\mu\nu}(p)_a = \frac{g_{\mu\nu} + \dfrac{p_\mu p_\nu}{m_A^2}}{p^2 + m_A^2 - i\varepsilon} + \frac{\dfrac{1}{m_A^2} p_\mu p_\nu}{p^2 + m_\pi^2 - i\varepsilon}.$$

(3.182)

All the longitudinal terms are ineffective in virtue of the propagation function $D_{\mu\nu}(p) [p^\mu D_{\mu\nu}(p) = 0]$, and therefore

$$G_{\mu\nu}(p)_\rho - G_{\mu\nu}(p)_a \rightarrow \frac{g_{\mu\nu}}{p^2 + m_\rho^2 - i\varepsilon} - \frac{g_{\mu\nu}}{p^2 + m_A^2 - i\varepsilon}.$$

(3.183)

But

$$g_{\mu\nu} D^{\mu\nu}(p) = \frac{3}{p^2 - i\varepsilon},$$

(3.184)

so that

$$\delta m_\pi^2 = 6e^2 m_\rho^2 \int \frac{1}{i} \frac{(dp)}{(2\pi)^4} \frac{1}{p^2 - i\varepsilon} \left[\frac{1}{p^2 + m_\rho^2 - i\varepsilon} \right.$$

$$\left. - \frac{1}{p^2 + m_A^2 - i\varepsilon} \right].$$

(3.185)

We now use Euclidean spherical coordinates $((1/i)dp_0 = dp_4)$

$$(1/i)(dp) \rightarrow \pi^2 p^2 dp^2,$$

and get

$$\delta m_\pi^2 = \frac{6e^2 m_\rho^2}{16\pi^2} \int_0^\infty dp^2 \left(\frac{1}{p^2 + m_\rho^2} - \frac{1}{p^2 + m_A^2} \right).$$

(3.186)

The result is

$$\delta m_\pi^2 = \left[\frac{3\alpha}{2\pi} \log 2 \right] m_\rho^2,$$

(3.187)

where the relation $m_A^2 = 2m_\rho^2$ has been used. This gives

$$\delta m_\pi = 5.0 \text{ MeV},$$

to be compared with the experimental value 4.6 MeV. The chirality calculation has several features. It is restricted to the pion. Implicit in this calculation is the neglect of the pion mass relative to those of other particles. The method seems to suggest that the particle A_1 is essential to a pion electromagnetic mass calculation, as indicated by the cancellation of the two propagation functions, $G_{\mu\nu}(p)_\rho$ and $G_{\mu\nu}(p)_a$.

We shall now describe briefly a more general approach to show that although the consideration of A_1 is significant, it is not fundamental. A not unsatisfactory result is obtained from the $\pi + \rho$ system alone. This method is not restricted to the π calculation. It recognizes that the action w_e produces an electromagnetic modification of the ρ^0 propagation function. The modification introduces new processes associated with ρ^0 exchange. Thus, for π mesons we have the couplings $\rho\pi\pi$ and $\pi A_1\rho$. Two π sources can exchange $\rho\pi$ or $A_1\rho$, as indicated in Fig. 3.3. These processes are already included in the

FIGURE 3.3

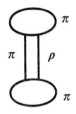

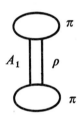

strong interaction phenomenology, but the electromagnetic modification effects are not. They produce new phenomena that can be evaluated individually. The results (for $m_\pi \ll m_\rho$) are

$$(\delta m_\pi^{\,2})_{\pi\pi\rho} = \frac{3\alpha}{4\pi} m_\rho^{\,2}, \quad (\delta m_\pi)_{\pi\pi\rho} = 3.6 \text{ MeV}. \qquad (3.188)$$

$$(\delta m_\pi^{\,2})_{\pi A_1\rho} = \frac{3\alpha}{4\pi} (2 \log 2 - 1) m_\rho^{\,2}, \quad (\delta m_\pi)_{\pi A_1\rho} = 1.4 \text{ MeV}. \qquad (3.189)$$

But in this approach we can also consider the $m_\pi^{\,2}$ correction to these processes. For the most important one $(\pi\pi\rho)$ we get

$$(\delta m_\pi^{\,2})_{\pi\pi\rho} = \frac{3\alpha}{4\pi} m_\rho^{\,2} \left[1 + \frac{m_\pi^{\,2}}{m_\rho^{\,2}} \left(\log \frac{m_\rho^{\,2}}{m_\pi^{\,2}} + \frac{1}{2} \right) \right] \qquad (3.190)$$

in which the second term gives a 13% correction, and $(\delta m_\pi)_{\pi\pi\rho}$ becomes

$$(\delta m_\pi)_{\pi\pi\rho} = 4.1 \text{ MeV}. \qquad (3.191)$$

The result is still qualitatively satisfactory, but the quantitative problem remains.

3.8 $U_4 \times U_4$ **Partial Symmetry**

We shall now discuss an entirely different example of partial
symmetry. Another aspect of the source formalism has not yet been mentioned. It is connected with the classification scheme of strongly interacting
particles, using the unitary group U_3 to label the particle multiplets. We use
the sources, or associated fields, to carry the indices. Let us designate the
two types of three-valued unitary indices by a and a^*. Then a source like
S_{ab*} can be decomposed into a unitary octuplet and a unitary singlet, 3×3
$= 8 + 1$. Similarly, the source S_{abc} which is totally symmetrical in abc
describes a unitary decuplet (10), and S_{abc} with an antisymmetrical pair bc
contains a unitary octuplet and a unitary singlet, since an antisymmetrical
pair of similar unitary indices is equivalent to a complex conjugate index.
If we employ the uniform representation of all spins by multispinors, the
complete structure of the source now takes the form

$$S_{\zeta_1 \ldots \zeta_n, \, a_1 \ldots a \, a^* \ldots a^*_{n'}} .$$

Particularly interesting is the subset with $n = n'$, which seems to describe
the best known particles. The source $S_{\zeta_1 a, \, \zeta_2 b*}$ describes spin-0 and spin-1
meson multiplets, each containing a unitary octuplet and singlet. This
source can be applied to the well-established 0^- and 1^- mesons. And the
source $S_{\zeta_1 a_1, \, \zeta_2 a_2, \, \zeta_3 a_3}$ which is totally symmetrical in the three pairs of
indices contains both spin $(1/2)$ and $(3/2)$, forming a unitary octuplet and
a decuplet, respectively. This source is applicable to the known system of
$(1/2)^+$ baryons and $(3/2)^+$ baryon resonances. These structures appear as
generalized sources which unite sources of individual particles with various
spins and a common parity, but with no implication about mass
degeneracy.

Let's consider the generalized meson source and for the moment omit the
unitary indices. The structure $S_{\zeta_1 \zeta_2}$ is a 4×4 matrix. This number of components is more than necessary to describe the spin multiplicities. The
physically significant components are selected by the use of projection
matrices. However, we can view this from a more general standpoint which
will give a connection with the field description we have adopted for spin-0
and spin-1 particles. The indices of the source $S_{\zeta_1 \zeta_2}$ can be classified by the
eigenvalues of γ^0 and the spin component σ along a certain direction, i.e.,
$\zeta = \gamma^0, \sigma$. Then we can decompose $S_{\zeta_1 \zeta_2}$ into four sets, in accordance with
eigenvalues of γ_1^0 and γ_2^0. This is a parity classification. The even-parity
components have $\gamma_1^{0\prime} = +1$, $\gamma_2^{0\prime} = +1$, or $\gamma_1^{0\prime} = -1$, $\gamma_2^{0\prime} = -1$. The
odd-parity components have $\gamma_1^{0\prime} = +1, \gamma_2^{0\prime} = -1$, or $\gamma_1^{0\prime} = -1, \gamma_2^{0\prime} = +1$.
The $(++)$ components form a matrix $S_{\sigma,\sigma'}^{(+)}$ and similarly $(--)$ com-

ponents form another matrix $S_{\sigma,\sigma'}{}^{(-)}$. In the Lagrange function describing the pion (spin 0, odd parity) and the ρ meson (spin 1, odd parity), the pion is represented by a pseudoscalar field π, and the ρ meson by a four-vector ρ^μ in which the spatial components ρ^k have odd parity, but the time component ρ^0 has even parity. In order to establish the correspondence between the multispinor description and the tensor field description, we use the equivalent first-order form for the Lagrange functions of π and ρ,

$$\mathscr{L}_\pi = -\pi^\mu \cdot \partial_\mu \pi + (1/2)(\pi_\mu)^2 - (1/2)m_\pi{}^2\pi^2, \tag{3.192}$$

and

$$\mathscr{L}_\rho = -(1/2)\rho^{\mu\nu} \cdot (\partial_\mu \rho_\nu - \partial_\nu \rho_\mu) + (1/4)(\rho_{\mu\nu})^2$$
$$-(1/2)m_\rho{}^2(\rho_\mu)^2. \tag{3.193}$$

The principle of stationary action implies the connections

$$\pi_\mu = \partial_\mu \pi, \tag{3.194}$$

$$\rho_{\mu\nu} = \partial_\mu \rho_\nu - \partial_\nu \rho_\mu, \tag{3.195}$$

and $\mathscr{L}_\pi$ and $\mathscr{L}_\rho$ restate the previous structures. But now the pion is described by five fields (π and π_μ), and the ρ meson by ten fields (ρ_μ and $\rho_{\mu\nu}$). Thus, on supplying one more scalar, we have a one to one correspondence between the multispinor and the tensor field description, each having 16 components.

Let's now consider a situation in which the meson fields carry arbitrary small spatial momenta,

$$\partial\chi = 0, \tag{3.196}$$

where χ is any meson field. Then we have

$$\mathscr{L}_\pi \to -\pi^0\partial_0\pi - (1/2)(\pi^0)^2 - (1/2)m_\pi{}^2\pi^2 \quad \text{odd-parity fields}$$
$$+ (1/2)(\pi_k)^2, \quad \text{even-parity fields} \tag{3.197}$$

and

$$\mathscr{L}_\rho \to -\rho^{0k}\partial_0\rho_k - (1/2)(\rho^{0k})^2$$
$$-(1/2)m_\rho{}^2(\rho_k)^2 \quad \text{odd-parity fields}$$
$$+ (1/2)H_\rho{}^2 + (1/2)m_\rho{}^2(\rho^0)^2, \quad \text{even-parity fields} \tag{3.198}$$

where $H_{\rho 1} = \rho_{23}$, etc. It is noted that the dynamical connections between odd-parity and even-parity fields are completely severed, in the absence of the spatial derivative terms, and new symmetries appear. If we restrict our attention to the U_2 subspace, the particles involved are $\pi, \eta_2(\eta, \eta^*), \rho$ and ω.

The correspondence between the multispinor notation and the tensor field description is established by introducing the 4×4 matrices

$$M^{(\pm)} = (\pi \pm H_\rho) \cdot \sigma\tau + (\eta_2 \pm H_\omega) \cdot \sigma$$
$$- (\pm S_1 + m_\rho \rho^0) \cdot \tau - (\pm S_0 + m_\omega \omega^0). \qquad (3.199)$$

These constructions are such that the even-parity structures in the Lagrange function are reproduced,

$$(1/16)\,\mathrm{Tr}\left[(M^{(+)})^2 + (M^{(-)})^2\right]$$
$$= (1/2)\left[(\pi)^2 + H_\rho^2 + m_\rho^2(\rho^0)^2 + H_\omega^2 + m_\omega^2(\omega^{02}) + ...\right]. \qquad (3.200)$$

In this form we recognize that this structure is invariant under unitary transformations on $M^{(+)}$ and $M^{(-)}$, independently. That is, we have a partial symmetry. It is the group $U_4 \times U_4$.

Next, consider the baryons in their rest frame. Since the momentum transfer is arbitrary small, the baryons retain a definite parity, and only even-parity meson field components contribute to the interaction. If one regards all the physical particles to be the outcome of some unknown dynamics, then even the symmetry or partial symmetry revealed by noninteracting particles has dynamical significance. We therefore make the partial symmetry hypothesis—that the meson-baryon coupling also possesses the U_4 symmetry. In the rest frame, the baryon field $\Psi_{\zeta_1 a_1, \zeta_2 a_2, \zeta_3 a_3}$ reduces effectively to $\Psi_{A_1 A_2 A_3}$ with $A = \sigma, \tau$, and ${\gamma_1}^{0\prime} = {\gamma_2}^{0\prime} = {\gamma_3}^{0\prime} = +1$. The latter is totally symmetrical in the four-valued indices A_1, A_2, and A_3. Such a coupling is

$$\frac{g}{2m_\rho}\, \Psi^* \sum_{\alpha=1}^{3} M_\alpha^{(+)} \Psi, \qquad (3.201)$$

where each $M_\alpha^{(+)}$ acts only on the corresponding indices. The scale factor is fixed by the requirement that ρ^0 couples universally to the isotopic-spin density $\Psi^* \sum_{\alpha=1}^{3} (1/2)\tau_\alpha \Psi$ giving the form

$$-g\rho^0 \cdot \Psi^* \sum_\alpha (1/2)\tau_\alpha \Psi. \qquad (3.202)$$

The nucleon part of the interaction (3.201) is

$$\frac{g}{2m_\rho} N^* \left[\frac{5}{3}(\pi + H_\rho) \cdot \sigma\tau + H_\omega \cdot \sigma - m_\rho \rho^0 \cdot \tau - 3m_\omega \omega^0 \right] N, \qquad (3.203)$$

where we have made use of the reductions for the nucleon

$$\sum_\alpha (1/2)\tau_\alpha \to (1/2)\tau,$$
$$\sum_\alpha (1/2)\sigma_\alpha \to (1/2)\sigma,$$
$$\sum_\alpha \sigma_\alpha \tau_\alpha \to (5/3)\sigma\tau, \qquad (3.204)$$

where the first two are simply the assertion that $(1/2)\tau$ and $(1/2)\sigma$ are the isotopic spin and ordinary spin of the nucleon, respectively. The third one is obtained by a simple calculation. The πN coupling

$$\frac{f}{m_\pi} \bar{N}\gamma^\mu i\gamma_5\tau N \cdot \partial_\mu\pi, \tag{3.205}$$

becomes, in the rest frame of the nucleon,

$$\frac{f}{m_\pi} N^*\sigma\tau N \cdot \pi. \tag{3.206}$$

Comparison of (3.203) with (3.206) implies the connection

$$\frac{f}{f_0} = \frac{(5/3)}{\sqrt{2}}$$

$$= 1.18\begin{cases} \text{Exp, strong interaction:} \dfrac{1.01\pm0.01}{0.84\pm0.03} = 1.20\pm0.03 \\[2mm] \text{Exp, weak interaction: } 1.18\pm0.03, \end{cases} \tag{3.207}$$

since

$$\frac{g}{2m_\rho} = \frac{1}{\sqrt{2}}\frac{f_0}{m_\pi}. \tag{3.208}$$

One also notices that the substitutions

$$\rho_{\mu3} \to \frac{e}{g} A_\mu,$$

$$\omega_\mu \to \frac{1}{3}\frac{m_\rho}{m_\omega}\frac{e}{g} A_\mu, \tag{3.209}$$

reproduce the correct electrical charges of the baryons. Thus (3.203) predicts the total magnetic moments for the nucleon

$$\mu = \frac{5}{3}\frac{e}{2m_\rho}\tau_3 + \frac{1}{3}\frac{e}{2m_\omega}. \tag{3.210}$$

If the mass difference between ω and ρ^0 is ignored, we recover the famous ratio

$$-\frac{\mu_\rho}{\mu_n} = \frac{3}{2} \qquad (m_\rho = m_\omega).$$

An assumed mass difference of 25 MeV gives the ratio 1.48. The experimental value is 1.46. The absolute moment predictions are about fifteen percent too small.

It is interesting to note the dynamical significance of the combination $\pi + H_\rho$ for nuclear forces. Consider a nonrelativistic two-nucleon system. The nuclear potential from these couplings is proportional to

$$\tau_1 \cdot \tau_2 \left[\sigma_1 \cdot \nabla \sigma_2 \cdot \nabla \frac{e^{-m_\pi r}}{r} + \sigma_1 \times \nabla \cdot \sigma_2 \times \nabla \frac{e^{-m_\rho r}}{r} \right]$$

$$= \tau_1 \cdot \tau_2 \left[\sigma_1 \cdot \nabla \sigma_2 \cdot \nabla \left(\frac{e^{-m_\pi r}}{r} - \frac{e^{-m_\rho r}}{r} \right) \right.$$

$$\left. + m_\rho^2 \sigma_1 \cdot \sigma_2 \frac{e^{-m_\rho r}}{r} \right].$$

When the two particles are close to each other, the π contribution alone involves a term $\sim (1/r^3)$, which is physically unacceptable. The particular combination $\pi + H_\rho$ removes the overly singular potential. The nucleon not only interacts with π but also with A_1 in the same manner. The much shorter ranged A_1 coupling tends to increase the tensor force of the π coupling. This suggests qualitively that the H_ρ coupling could be stronger than indicated. If so, the predictions about the absolute magnetic moments for the nucleon will then be improved.

3.9   Concluding Remarks

As a phenomenological description the source theory has reproduced and even improved, without unnecessary speculative assumptions, the successful results of other formalisms. As a computational tool it has also simplified many of the calculations. I hope in these lectures I have given you enough of the general view, the method, and the spirit of the source theory that you will think seriously of accepting the source theory for what it is—a self-contained comprehensive approach to particle physics, which is intended to replace the present competing methods by a simpler synthesis.

These notes record the state of the art of Sourcery in July, 1967. For more details and later developments, go to the Source papers published in the Physical Review. They are:

"Particles and Sources": **152**, 1219 (1966);
"Sources and Electrodynamics": **158**, 1391 (1967);
"Gauge Fields, Sources and Electromagnetic Masses": **165**, 1714 (1968); **167**, 1546 (1968);
"Sources and Gravitons": **173**, 1264 (1968);
"Sources and Magnetic Charge": **173**, 1536 (1968).

See also:

"Chiral Transformations": **167**, 1432 (1968)
and these Physical Review Letters:
"Partial Symmetry": **18**, 923 (1967);
"Photons, Mesons and Form Factors": **19**, 1154 (1967);
"Radiative Corrections in β Decay": **19**, 1501 (1967).